EXPERIMENTAL DESIGNS

THE SAGE QUANTITATIVE RESEARCH KIT

Experimental Designs by *Barak Ariel, Matthew Bland* and *Alex Sutherland* is the 6th volume in *The SAGE Quantitative Research Kit*. This book can be used together with the other titles in the *Kit* as a comprehensive guide to the process of doing quantitative research, but is equally valuable on its own as a practical introduction to the various types of experimental designs available to researchers today.

Editors of The SAGE Quantitative Research Kit:

Malcolm Williams – *Cardiff University, UK*

Richard D. Wiggins – *UCL Social Research Institute, UK*

D. Betsy McCoach – *University of Connecticut, USA*

Founding editor:

The late W. Paul Vogt – *Illinois State University, USA*

EXPERIMENTAL DESIGNS

BARAK ARIEL
MATTHEW BLAND
ALEX SUTHERLAND

Los Angeles | London | New Delhi
Singapore | Washington DC | Melbourne

THE SAGE QUANTITATIVE RESEARCH KIT

Los Angeles | London | New Delhi
Singapore | Washington DC | Melbourne

SAGE Publications Ltd
1 Oliver's Yard
55 City Road
London EC1Y 1SP

SAGE Publications Inc.
2455 Teller Road
Thousand Oaks, California 91320

SAGE Publications India Pvt Ltd
B 1/I 1 Mohan Cooperative Industrial Area
Mathura Road
New Delhi 110 044

SAGE Publications Asia-Pacific Pte Ltd
3 Church Street
#10-04 Samsung Hub
Singapore 049483

© Barak Ariel, Matthew Bland and Alex Sutherland 2021

This volume published as part of *The SAGE Quantitative Research Kit* (2021), edited by Malcolm Williams, Richard D. Wiggins and D. Betsy McCoach.

Editor: Jai Seaman
Assistant editor: Charlotte Bush
Production editor: Manmeet Kaur Tura
Copyeditor: QuADS Prepress Pvt Ltd
Proofreader: Elaine Leek
Indexer: Caroline Eley
Marketing manager: Susheel Gokarakonda
Cover design: Shaun Mercier
Typeset by: C&M Digitals (P) Ltd, Chennai, India
Printed in the UK

Library of Congress Control Number: 2020949755

British Library Cataloguing in Publication data

A catalogue record for this book is available from the British Library

ISBN 978-1-5264-2662-8

At SAGE we take sustainability seriously. Most of our products are printed in the UK using responsibly sourced papers and boards. When we print overseas, we ensure sustainable papers are used as measured by the PREPS grading system. We undertake an annual audit to monitor our sustainability.

The word *experiment* is used in a quite precise sense to mean an investigation where the system under study is under the control of the investigator.

    – David Cox and Nancy Reid (2000, p. 1). *The Theory of the Design of Experiments.*

CONTENTS

LIST OF FIGURES, TABLES AND BOXES

List of figures

List of tables

List of boxes

List of case studies

ABOUT THE AUTHORS

Barak Ariel is a Reader in experimental criminology at the Institute of Criminology, University of Cambridge, and an Associate Professor at the Institute of Criminology in the Faculty of Law, Hebrew University of Jerusalem. He is a fellow at the Jerry Lee Centre of Experimental Criminology and at the Division of Experimental Criminology of the American Society of Criminology, which he presently chairs. He has been involved in dozens of experiments representing a wide range of issues in law enforcement, including organised crime, offender management, counterterrorism, legitimacy, neighbourhood policing, restorative justice, domestic violence, crime and place, body-worn cameras, Tasers and the broader use of technology. Much of this body of work appears in more than 100 academic articles and books. Ariel sits on multiple evaluation panels and has served as an adviser to policing and security agencies around the globe, reviewing their impact evaluations and evidence-based practices largely through randomised controlled trials.

Matthew Bland is a Lecturer in evidence-based policing at the Institute of Criminology, University of Cambridge, and a Research Fellow at the Cambridge Centre for Evidence-Based Policing. At the university, he teaches senior police leaders through the Police Executive Programme. He has managed and contributed to experiments on offender management, hotspot policing and out-of-court disposals. His research focuses on domestic abuse, prediction algorithms in policing and exploratory analyses. He previously worked in policing for 15 years as a Head of Analysis, during which time he completed both a master's and a PhD in criminology at the University of Cambridge.

Alex Sutherland is the Chief Scientist and Director of Research and Evaluation for the Behavioural Insights Team. He has nearly 20 years of experience in experimental designs and evaluation, including work on the use and understanding of research evidence for decision-making and policy development. His research interests lie in criminal justice, violence and violence prevention. He previously worked at RAND Europe and the University of Cambridge. He is currently a member of the UK Government Trial Advice Panel.

1

INTRODUCTION

Chapter Overview

Formal textbooks on experiments first surfaced more than a century ago, and thousands have emerged since then. In the field of education, William McCall published *How to Experiment in Education* in 1923; R.A. Fisher, a Cambridge scholar, released *Statistical Methods for Research Workers* and *The Design of Experiments* in 1925 and 1935, respectively; S.S. Stevens circulated his *Handbook of Experimental Psychology* in 1951. We also have D.T. Campbell and Stanley's (1963) classic *Experimental and Quasi-Experimental Designs for Research*, and primers like Shadish et al.'s (2002) *Experimental and Quasi-Experimental Designs for Generalised Causal Inference*, which has been cited nearly 50,000 times. These foundational texts provide straightforward models for using experiments in causal research within the social sciences.

Fundamentally, this corpus of knowledge shares a common long-standing methodological theme: when researchers want to attribute **causal inferences** between interventions and outcomes, they need to conduct experiments. The basic model for demonstrating cause-and-effect relationships relies on a formal, scientific process of hypothesis testing, and this process is confirmed through the experimental design. One of these fundamental processes dictates that causal inference necessarily requires a *comparison*. A valid test of any intervention involves a situation through which the treated group (or units) can be compared – what is termed a **counterfactual**. Put another way, evidence of 'successful treatment' is always relative to a world in which the treatment was *not* given (D.T. Campbell, 1969). Whether the treatment group is compared to itself prior to the exposure to the intervention, or a separate group of cases unexposed to the intervention, or even just some predefined criterion (like a national average or median), contrast is needed. While others might disagree (e.g. Pearl, 2019), without an objective comparison, we cannot talk about causation.

Causation theories are found in different schools of thought (for discussions, see Cartwright & Hardie, 2012; Pearl, 2019; Wikström, 2010). The dominant causal framework is that of 'potential outcomes' (or the Neyman–Rubin causal framework; Rubin, 2005), which we discuss herein and which many of the designs and examples in this book use as their basis. Until mainstream experimental disciplines revise the core foundations of the standard scientific inquiry, one must be cautious when recommending public policy based on alternative research designs. Methodologies based on subjective or other schools of thought about what **causality** means will not be discussed in this book. To emphasise, we do not discount these methodologies and their contribution to research, not least for developing logical hypotheses about the causal relationships in the universe. We are, however, concerned about risks to the validity of these causal claims and how well they might stand a chance of being implemented in practice. We discuss these issues in more detail in Chapter 4. For further reading, see Abell and Engel (2019) as well as Abend et al. (2013).

However, not all comparisons can be evaluated equally. For the inference that a policy or change was 'effective', researchers need to be sure that the comparison group that was not exposed to the intervention resembles the group that was exposed to the intervention as much as possible. If the treatment group and the no-treatment group are incomparable – not 'apples to apples' – it then becomes very difficult to 'single out' the treatment effect from pre-existing differences. That is, if two groups differ before an intervention starts, how can we be sure that it was the introduction of the intervention and not the pre-existing differences that produce the result?

To have confidence in the conclusions we draw from studies that look at the causal relationship between interventions and their outcomes means having only one attributable difference between treatment and no-treatment conditions: the treatment itself. Failing this requirement suggests that any observed difference between the treatment and no-treatment groups can be attributed to other explanations. Rival hypotheses (and evidence) can then falsify – or confound – the hypothesis about the causal relationship. In other words, if the two groups are not comparable at baseline, then it can be reasonably argued that the outcome was caused by inherent differences between the two groups of **participants**, by discrete settings in which data on the two groups were collected, or through diverse ways in which eligible cases were recruited into the groups. Collectively, these plausible yet alternative explanations to the observed outcome, other than the treatment effect, undermine the test. Therefore, a reasonable degree of 'pre-experimental comparability' between the two groups is needed, or else the claim of causality becomes speculative. We spend a considerable amount of attention on this issue throughout the book, as all experimenters share this fundamental concern regarding equivalence.

Experiments are then split into two distinct approaches to achieve pre-experimental comparability: *statistical designs* and **randomisation**. Both aim to facilitate equitable conditions between treatment and control conditions but achieve this goal differently. Statistical designs, often referred to as *quasi-experimental* methods, rely on statistical analysis to control and create equivalence between the two groups. For example, in a study on the effect of police presence on crime in particular neighbourhoods, researchers can compare the crime data in 'treatment neighbourhoods' before and after patrols were conducted, and then compare the results with data from 'control neighbourhoods' that were not exposed to the patrols (e.g. Kelling et al., 1974; Sherman & Weisburd, 1995). Noticeable differences in the before–after comparisons would then be attributed to the police patrols. However, if there are also observable differences between the neighbourhoods or the populations who live in the treatment and the no-treatment neighbourhoods, or the types of crimes that take place in these neighbourhoods, we can use statistical controls to 'rebalance' the groups – or at least account for the differences between groups arising from these other variables.

Through statistically controlling for these other variables (e.g. Piza & O'Hara, 2014; R.G. Santos & Santos, 2015; see also *The SAGE Quantitative Research Kit*, Volume 7), scholars could then match patrol and no-patrol areas and take into account the confounding effect of these other factors. In doing so, researchers are explicitly or implicitly saying 'this is as good as randomisation'. But what does that mean in practice?

While on the one hand, we have statistical designs, on the other, we have experiments that use randomisation, which relies on the mathematical foundations of probability theory (as discussed in *The SAGE Quantitative Research Kit*, Volume 3). Probability theory postulates that through the process of randomly assigning cases into treatment and no-treatment conditions, experimenters have the best shot of achieving pre-experimental comparability between the two groups. This is owing to the law of large numbers (or 'logic of science' according to Jaynes, 2003). Allocating units at random does, with a large enough sample, create balanced groups. As we illustrate in Chapter 2, this balance is not just apparent for observed variables (i.e. what we can measure) but also in terms of the unobserved factors that we cannot measure (cf. Cowen & Cartwright, 2019). For example, we can match treatment and comparison neighbourhoods in terms of crimes reported to the police before the intervention (patrols), and then create balance in terms of this variable (Saunders et al., 2015; see also Weisburd et al., 2018). However, we cannot create true balance between the two groups if we do not have data on *un*reported crimes, which may be very different in the two neighbourhoods.

We cannot use statistical controls where no data exist or where we do not measure something. The randomisation of units into treatment and control conditions largely mitigates this issue (Farrington, 2003a; Shadish et al., 2002; Weisburd, 2005). This quality makes, in the eyes of many, randomised experiments a superior approach to other designs when it comes to making causal claims (see the debates about 'gold standard' research in Saunders et al., 2016). Randomised experiments have what is called a high level of *internal validity* (see review in Grimshaw et al., 2000; Schweizer et al., 2016). What this means is that, when properly conducted, a randomised experiment gives one the greatest confidence levels that the effect(s) observed arose because of the cause (randomly) introduced by the experiment, and not due to something else.

The parallel phrase – ***external validity*** – means the extent to which the results from this experiment can apply elsewhere in the world. Lab-based randomised experiments typically have very high internal validity, but very low external validity, because their conditions are highly regulated and not replicable in a 'real-world' scenario. We review these issues in Chapter 3.

Importantly, random allocation means that randomised experiments are prospective not retrospective – that is, testing forthcoming interventions, rather than ones that have already been administered where data have already been produced. Prospective studies allow researchers to maintain more control compared to retrospective studies.

The researcher is involved in the very process of case selection, treatment fidelity (the extent to which a treatment is delivered or implemented as intended) and the data collated for the purposes of the experiment. Experimenters using random assignment are therefore involved in the distribution and management of units into different real-life conditions (e.g. police patrols) *ex ante* and not *ex post*. As the scholar collaborates with a treatment provider to jointly follow up on cases, and observe variations in the measures within the treatment and no-treatment conditions, they are in a much better position to provide assurance that the fidelity of the test is maintained throughout the process (Strang, 2012). These features rarely exist in quasi-experimental designs, but at the same time, randomised experiments require scientists to pay attention to maintaining the proper controls over the administration of the test. For this reason, running a randomised controlled trial (RCT) can be laborious.

In Chapter 5, we cover an underutilised instrument – the experimental protocol – and illustrate the importance of conducting a pre-mortem analysis: designing and crafting the study before venturing out into the field. The experimental protocol requires the researcher to address ethical considerations: how we can secure the rights of the participants, while advancing scientific knowledge through interventions that might violate these rights. For example, in policing experiments where the participants are offenders or victims, they do not have the right to consent; the policing strategy applied in their case is predetermined, as offenders may be mandated by a court to attend *a* treatment for domestic violence. However, the allocation of the offenders into any *specific* treatment is conducted randomly (see Mills et al., 2019). Of course, if we *know* that a particular treatment yields better results than the comparison treatment (e.g. reduces rates of repeat offending compared to the rates of reoffending under control conditions), then there is no ethical justification for conducting the experiment. When we do not have evidence that supports the hypothesised benefit of the intervention, however, then it is unethical *not* to conduct an experiment. After all, the existing intervention for domestic batterers can cause backfiring effects and lead to more abuse. This is where experiments are useful: they provide evidence on *relative* utility, based on which we can make sound policy recommendations. Taking these points into consideration, the researcher has a duty to minimise these and other ethical risks as much as possible through a detailed plan that forms part of the research documentation portfolio.

Vitally, the decision to randomise must also then be followed with the question of which 'units' are the most appropriate for random allocation. This is not an easy question to answer because there are multiple options, thus the choice is not purely theoretical but a pragmatic query. The decision is shaped by the very nature of the field, settings and previous tests of the intervention. Some units are more suitable for addressing certain theoretical questions than others, so the size of the study matters, as well as the dosage of the treatment. Data availability and feasibility also determine these choices. Experimenters need to then consider a wide range of methods of actually conducting the random assignment, choosing between simple, **'trickle flow'**, block random assignment, cluster, stratification and other perhaps more nuanced and bespoke sequences of random allocation designs. We review each of these design options in Chapter 2.

We then discuss issues with control with some detail in Chapter 3. The mechanisms used to administer randomised experiments are broad, and the technical literature on these matters is rich. Issues of group imbalances, sample sizes and measurement considerations are all closely linked to an unbiased experiment. Considerations of these problems begin in the planning stage, with a pre-mortem assessment of the possible pitfalls that can lead the experimenter to lose control over the test (see Klein, 2011). Researchers need to be aware of threats to internal validity, as well as the external validity of the experimental tests, and find ways to avoid them during the experimental cycle. We turn to these concerns in Chapter 3 as well.

In Chapter 4, we account for the different types of experimental designs available in the social sciences. Some are as 'simple' as following up with a group of participants after their exposure to a given treatment, having been randomly assigned into treatment and control conditions, while others are more elaborate, multistage and complex. The choice of applying one type of test and not another is both conceptual and pragmatic. We rely heavily on classic texts by D.T. Campbell and Stanley (1963), Cook and Campbell (1979) and the amalgamation of these works by Shadish et al. (2002), which detail the mechanics of experimental designs, in addition to their rationales and pitfalls. However, we provide more updated examples of experiments that have applied these designs within the social sciences. Many of our examples are criminological, given our backgrounds, but are applicable to other experimental disciplines.

Chapter 4 also provides some common types of quasi-experimental designs that can be used when the conditions are not conducive to random assignment (see Shadish et al., 2002, pp. 269–278). Admittedly, the stack of evidence in causal research largely comprises statistical techniques, including the regression discontinuity design, propensity score **matching**, difference-in-difference design, and many others. We introduce these approaches and refer the reader to the technical literature on how to estimate causal inference with these advanced statistics.

Before venturing further, we need to contextualise experiments in a wide range of study designs. Understanding the role that causal research has in science, and what differentiates it from other methodological approaches, is a critical first step. To be clear, we do not argue that experiments are 'superior' compared to other methods; put simply, the appropriate research design follows the research question and the research settings. The utility of experiments is found in their ability to allow researchers to test specific hypotheses about causal relationships. Scholars interested in longitudinal processes, qualitative internal dynamics (e.g. perceptions) or descriptive assessments of phenomena use observational designs. These designs are a good fit for these lines of scientific inquiries. Experiments – and within this category we include both quasi-experimental designs and RCTs of various types – are appropriate when making causal inferences.

Finally, we then defend the view that precisely the same arguments can be made by policymakers who are interested in **evidence-based policy**: experiments are needed for impact evaluations, preferably with a randomisation component of allocating cases into treatment(s) and tight controls over the **implementation** of the study design. We discuss these issues in the final chapter, when we speculate more about the link between experimental evidence and policy.

Contextualising randomised experiments in a wide range of causal designs

RCTs are (mostly) regarded as the 'gold standard' of impact evaluation research (Sherman et al., 1998). The primary reason for this affirmation is **internal validity**, which is the feature of a test that tells us that it measures what it claims to measure (Kelley, 1927, p. 14). Simply put, well-designed randomised experiments that are correctly executed have the highest possible internal validity to the extent that they enable the researcher to quantifiably demonstrate that a variation in a treatment (what we call changes in the '**independent variable**') causes variation(s) in an outcome, or the '**dependent variable**(s)' (Cook & Campbell, 1979; Shadish et al., 2002). We will contextualise randomised experiments against other causal designs – this is more of a level playing field – but then illustrate that 'basically, statistical control is not as good as experimental control' (Farrington, 2003b, p. 219) and 'design trumps analysis' (Rubin, 2008, p. 808).

Another advantage of randomised experiments is that they account for what is called *selection bias* – that is, results derived from choices that have been made or selection processes that create differences – artefacts of selection rather than true differences between treatment groups. In non-randomised controlled designs, the treatment group is selected on the basis of its success, meaning that the treatment provider has an inherent interest to recruit members who would benefit from it. This is natural, as the interest of the treatment provider is to assist the participants with what they believe is an effective intervention. Usually, patients with the best prognosis are participants who express the most desire to improve their situation, or individuals who are the most motivated to successfully complete the intervention programme. As importantly, the participants themselves often chose if and how to take part in the treatment. They have to engage, follow the treatment protocol and report to a data collector. By implication, this selection 'leaves behind' individuals who do not share these qualities even if they come from the same cohort or have similar characteristics (e.g. criminal history, educational background or sets of relevant skills). In doing so, the treatment provider gives an unfair edge to the treatment group over the comparison group: they are, by definition of this process, more likely to excel.[1]

The **bias** can come in the allocation process. Treatment providers might choose those who are more motivated, or who they think will be successful. Particularly if the selection process is not well documented, it is unsurprising that the **effect size** (the magnitude of the difference between treatment and **control groups** following the intervention) is larger than in studies in which the allocation of the cases into treatment and control conditions is conducted impartially. Only under these latter conditions can we say that the treatment has an equal opportunity to 'succeed' or 'fail'. Moreover, under ideal scenarios, even the researchers would be unaware of whom they are allocating to treatment and control conditions, thus 'blinding' them from intentionally or unintentionally allocating participants into one or the other group (see Day & Altman, 2000). In a 'blinded' random distribution, the fairest allocation is maintained. Selection bias is more difficult to avoid in non-randomised designs. In fact, matching procedures in field settings have led at least one synthesis of evidence (on the backfiring effect of participating in Alcoholics Anonymous programmes) to conclude that 'selection biases compromised all **quasi-experiments**' (Kownacki & Shadish, 1999).

Randomised experiments can also address the ***specification error*** encountered in observational models (see Heckman, 1979). This error term refers to the impossibility of including all – if not most – of the detrimental factors affecting the dependent variable studied. Random assignment of 'one condition to half of a large population by a formula that makes it equally likely that each subject will receive one treatment or another' generates comparable distribution in each of the two groups of factors 'that could affect results' (Sherman, 2003, p. 11). Therefore, the most effective way to study crime and crime-related policy is to intervene in a way that will permit the researcher to make a valid assessment of the intervention effect. A decision-making process that

---

[1]Notably, however, researchers resort to quasi-experimental designs especially when policies have been rolled out without regard to evaluation, and the fact that some cases were 'creamed in' is not necessarily borne out of an attempt to cheat. Often, interventions are simply put in place with the primary motivation of helping those who would benefit the most from the treatment. This means that we should not discount quasi-experimental designs, but rather accept their conclusions with the necessary caveats.

relies on randomised experiments will result in more precise and reliable answers to questions about what works for policy and practice decision-makers.

In light of these (and other) advantages of randomised experiments, it might be expected that they would be widely used to investigate the causes of offending and the **effectiveness** of interventions designed to reduce offending. However, this is not the case. Randomised experiments in criminology and criminal justice are relatively uncommon (Ariel, 2009; Farrington, 1983; Weisburd, 2000; Weisburd et al., 1993; see more recently Dezember et al., 2020; Neyroud, 2017), at least when compared to other disciplines, such as psychology, education, engineering or medicine. We will return to this scarcity later on; however, for now we return to David Farrington:

> The history of the use of randomised experiments in criminology consists of feast and famine periods . . . in a desert of nonrandomised research. (Farrington, 2003b, p. 219)

We illustrate more thoroughly why this is the case and emphasise why and how we should see more of these designs – especially given criminologists' focus on 'what works' (Sherman et al., 1998), and the very fact that **efficacy** and utility are best tested using experimental rather than non-experimental designs. Thus, in Chapter 6, we will also continue to emphasise that not all studies in criminal justice research can, or should, follow the randomised experiments route. When embarking on an impact evaluation study, researchers should choose the most fitting and cost-effective approach to answering the research question. This dilemma is less concerned with the substantive area of research – although it may serve as a good starting point to reflect on past experiences – and more concerned with the ways in which such a dilemma can be answered empirically and structurally.

Causal designs and the scientific meaning of causality

Causality in science means something quite specific, and scholars are usually in agreement about three minimal preconditions for declaring that a causal relationship exists between cause(s) and effect(s):

1   That there is a correlation between the two variables.
2   That there is a temporal sequence, whereby the assumed cause precedes the effect.
3   That there are no alternative explanations.

Beyond these criteria, which date back as far as the 18th-century philosopher David Hume, others have since added the requirement (4) for a *causal mechanism* to be explicated (Congdon et al., 2017; Hedström, 2005); however, more crucially in the context of policy evaluation, there has to be some way of *manipulating* the cause (for a more elaborate discussion, see Lewis, 1974; and the premier collection of papers on causality edited by Beebee et al., 2009). As clearly laid out by Wikström (2008),

> If we cannot manipulate the putative cause/s and observe the effect/s, we are stuck with analysing patterns of association (correlation) between our hypothesised causes and effects. The question is then whether we can establish causation (causal dependencies) by analysing patterns of association with statistical methods. The simple answer to this question is most likely to be a disappointing 'no'. (p. 128)

Holland (1986) has the strictest version of this idea, which is often paraphrased as 'no causation without manipulation'. That in turn has spawned numerous debates on the manipulability of causes being a prerequisite for causal explanation. As Pearl (2010) argues, however, *causal explanation* is a different endeavour.

Taking the three prerequisites for determining causality into account, it immediately becomes clear why observational studies are not in a position to prove causality. For example, Tankebe's (2009) research on legitimacy is valuable for indicating the relative role of procedural justice in affecting the community's sense of police legitimacy. However, this type of research cannot firmly place procedural justice as a causal antecedent to legitimacy because the chronological ordering of the two variables is difficult to lay out within the constraints of a cross-sectional survey.

Similarly, *one-group* longitudinal studies have shown significant (and negative) correlations between age and criminal behaviour (Farrington, 1986; Hirschi & Gottfredson, 1983; Sweeten et al., 2013).[2] In this design, one group of participants is followed over a period of time to illustrate how criminal behaviour fluctuates across different age brackets. The asymmetrical, bell-shaped age–crime curve illustrates that the proportion of individuals who offend increases through adolescence, peaks around the ages of 17 to 19, and then declines in the early 20s (Loeber & Farrington, 2014). For example, scholars can study a cohort of several hundred juvenile delinquents released from a particular institution between the 1960s and today, and learn when they committed offences to assess whether they exhibit the same age–crime curve. However, there is no attempt to compare their behaviour to any other group of participants. While we can show there is a link between the age of the offender and the number of crimes they committed over a life course, we cannot argue that age *causes* crime. Age 'masks' the causal factors that are *associated* with these age brackets (e.g. peer influence, bio-socio-psychological factors, strain). Thus, this line of **observational research** can firmly illustrate the temporal sequence of crime over time, but it cannot sufficiently rule out *alternative* explanations (outside of the age factor)

---

[2]We note the distinction between different longitudinal designs that are often incorrectly referred to as a single type of research methodology. We discuss these in Chapter 4.

to the link between age and crime (Gottfredson & Hirschi, 1987). Thus, we ought to be careful in concluding causality from observational studies.[3]

Even in more complicated, group-based trajectory analyses, establishing causality is tricky. These designs are integral to showing how certain clusters of cases or offenders change over time (Haviland et al., 2007). For instance, they can convincingly illustrate how people are clustered based on the frequency or severity of their offending over time. They may also use available data to control for various factors, like ethnicity or other socio-economic factors. However, as we discussed earlier, they suffer from the *specification error* (see Heckman, 1979): there may be *more* variables that explain crime *better* than grouping criterion (e.g. resilience, social bonds and internal control mechanisms, to name a few), which often go unrecorded and therefore cannot be controlled for in the statistical model.

Why should governments and agencies care about causal designs?

Criminology, especially policing research, is an applied science (Bottoms & Tonry, 2013). It therefore offers a case study of a long-standing discipline that directly connects academics and experimentalists with treatment providers and policymakers. This is where evidence-based practice comes into play: when practitioners use scientific evidence to guide policy and practice. Therefore, our field provides insight for others in the social sciences who may aspire towards more robust empirical foundations for applying tested strategies in real-life conditions.

Admittedly, RCTs remain a small percentage of studies in many fields, including criminology (Ariel, 2011; Dezember et al., 2020). However, educators, or psychologists, or nurses do not always follow the most rigorous research evidence when interacting with members of the public (Brants-Sabo & Ariel, 2020). Even physicians suffer from the same issues, though to a lesser extent (Grol, 2001). So while there is generally wide agreement that governmental branches should ground their decisions (at least in part) on the best data available, or, at the very least, evidence that supports a policy (Weisburd, 2003), there is still more work to be done before the symbiotic relationship between research and industry – that is, between science and practice – matures similarly to its development in the field of medicine.

Some change, at least in criminology, has been occurring in more recent years (see Farrington & Welsh, 2013). Governmental agencies that are responsible for upholding

---

[3]On the question of causality, see Cartwright (2004), but also see the excellent reply in Casini (2012).

the law rely more and more on research evidence to shape public policies, rather than experience alone. When deciding to implement interventions that 'work', there is a growing interest in evidence produced through rigorous studies, with a focus on RCTs rather than on other research designs. In many situations, policies have been advocated on the basis of ideology, pseudo-scientific methodologies and general conditions of ineffectiveness. In other words, such policies were simply not evidence-based approaches, ones that are not established on systematic observations (Welsh & Farrington, 2001).

Consequently, we have seen a move towards more systematic evaluations of crime-control practices in particular, and public policies in general, imbuing these with a scientific research base. This change is part of a more general movement in other disciplines, such as education (Davies, 1999; Fitz-Gibbon, 1999; Handelsman et al., 2004), psychology (among many others, see Webley et al., 2001), economics (Alm, 1991) and medicine. As an example, the Cochrane Library has approximately 2000 evidence-based medical and healthcare studies, and is considered the best singular source of such studies. This much-needed vogue in crime prevention policy began attracting attention some 15 years ago due to either 'growing pragmatism or pressures for accountability on how public funds are spent' (Petrosino et al., 2001, p. 16). Whatever the reason, evidence-based crime policy is characterised by 'feast and famine periods' as Farrington puts it, which are influenced by either key individuals (Farrington, 2003b) or structural and cultural factors (Shepherd, 2003). 'An evidence-based approach', it was said, 'requires that the results of rigorous evaluation be rationally integrated into decisions about interventions by policymakers and practitioners alike' (Petrosino, 2000, p. 635). Otherwise, we face the peril of implementing evidence-misled policies (Sherman, 2001, 2009).

The aforementioned suggests that there is actually a moral imperative for conducting randomised controlled experiments in field settings (see Welsh & Farrington, 2012). This responsibility is rooted in researchers' obligation to rely on empirical and compelling evidence when setting practices, policies and various treatments in crime and criminal justice (Weisburd, 2000, 2003). For example, the Campbell Collaboration Crime and Justice Group, a global network of practitioners, researchers and policymakers in the field of criminology, was established to 'prepare **systematic reviews** of high-quality research on the effects of criminological intervention' (Farrington & Petrosino, 2001, pp. 39–42). Moreover, other local attempts have provided policymakers with experimental results as well (Braithwaite & Makkai, 1994; Dittmann, 2004; R.D. Schwartz & Orleans, 1967; Weisburd & Eck, 2004). In sum, randomised experimental studies are considered one of the better ways to assess intervention effectiveness in criminology as part of an overall evidence-led policy imperative in public services (Feder & Boruch, 2000; Weisburd & Taxman, 2000; Welsh & Farrington, 2001; however cf. Nagin & Sampson, 2019).

Chapter Summary

- What is meant by employing an experiment as the research method? What are randomised controlled trials (RCTs) and how are they different from other kinds of controlled experiments that seek to produce causal estimates? Why is randomisation considered by many to be the 'gold standard' of evaluation research? What are the components of the R–C–Ts (random–control–trial), in pragmatic terms? This book highlights the importance of experiments and randomisation in particular for evaluation research, and the necessary controls needed to produce valid causal estimates of treatment effects.
- We review the primary experimental designs that can be used to test the effectiveness of interventions in social and health sciences, using illustrations from our field: criminology. This introductory chapter summarises these concepts and lays out the roadmap for the overall book.

Further Reading

Ariel, B. (2018). Not all evidence is created equal: On the importance of matching research questions with research methods in evidence-based policing. In R. Mitchell & L. Huey (Eds.), *Evidence-based policing: An introduction* (pp. 63–86). Policy Press.

This chapter provides further reading on the position of causal designs within research methods from a wider perspective. It lays out the terrain of research methods and provides a guide on how to select the most appropriate research method for different types of research questions.

Sherman, L. W. (1998). *Evidence-based policing*. The Police Foundation.

Sherman, L. W. (2013). The rise of evidence-based policing: Targeting, testing, and tracking. *Crime and justice, 42*(1), 377–451.

Evidence-based policy, or the use of scientific evidence to implement guidelines and evaluate interventions, has gained traction in different fields. In criminology, the scholar who has made the most profound contribution to 'evidence-based policing' is Professor Lawrence Sherman. On this topic, these two equally important papers should be consulted: Sherman (1998) systematically introduces a paradigm for evidence-based policing; and in Sherman (2013) the composition of evidence-based policing is laid out under the 'triple-T' strategy: targeting, testing and tracking.

2

R IS FOR RANDOM

Chapter Overview

Comparisons in science

As we noted in Chapter 1, causal inference requires a comparison, a benchmark or evidence on what would have happened if the policy were not in place. For example, when a claim is made about 'reducing crime' or 'improving well-being', one should ask oneself, 'What is the effect of this policy, compared to *not* having the policy, or compared to a *different* policy?' To claim that a policy or an intervention works necessitates an alternative, or comparator. To illustrate what a comparator is, consider the following example. Suppose you wish to test the effect of participating in Alcoholics Anonymous (AA) on drinking (see the review of studies in Kownacki & Shadish, 1999). You may *assume* that participation causes a reduction in excessive drinking, because it is a logical claim. However, how do you *know* that AA alone causes the change? What type of systematic evidence would convince you that AA 'works'?

One approach is to compare the participants' current drinking habits to their drinking habits before they joined AA (e.g. calculating the number of drinks per week). This 'before–after' analysis – that is, the same group measured before exposure to the intervention (in this case, AA sessions) and then again after the exposure – is a common design in evaluation research.

However, a closer look will tell us that this is a weak methodology for establishing causality: there are many alternative reasons for having more or less drinks per week in the 'after' period, *beyond* the AA intervention. We do not know whether those people would have reduced their drinking had they not taken part in the AA group. We cannot ascertain whether some feature in the desire to participate caused the change in drinking behaviour and perhaps an alternative alcohol treatment would be more effective than AA. With a before–after design, we cannot even rule out that something else happened to coincide with their changing behaviour. It might have been a change in the price of alcohol or the change in the weather. Being confined inside during the COVID-19 lockdown may have also contributed in some way.

To complicate the matter further, even if there is a pronounced difference between the 'before' and the 'after' periods, the researcher still cannot rule out an alternative (logical) reason for the difference. Extreme statistical fluctuations over time happen regularly in nature as well as in human behaviour. We provide a more technical reason why such changes might occur when we discuss chapter 3 the concept *regression to the mean*, but the fact that we have observed a large variation, especially in time-limited series of data, is unsurprising.

Collectively, one should see why we cannot conclusively argue that the change in drinking behaviour is *because* of AA using a before–after design. The causal link claimed between AA and reduced drinking would remain weak under these study conditions. It is so weak, in fact, that most scientists would dismiss a claim based on this before–after study design (Knapp, 2016). What we need then is a group to compare to our participants, what is called a counterfactual group. They need to be as similar as possible to our AA group participants – with the only difference being that they do *not* go to the AA sessions. So, how can we do this?

Box 2.1

When Is a Simple Before–After Design Enough for Inferring Causality?

There are research questions that do not require a design that is more sophisticated than the before–after research design. When trying to answer questions about an intervention so potent that there simply is no alternative explanation to the link between the intervention and the outcome, a before–after design is enough. For example, the effect of deportation on violent criminal behaviour of the deportee in the state from which they were extradited does not usually require anything beyond a before and after design; once they are gone, they cannot commit more offences in the home state, so we do not need a comparison to explain this. The effect on recidivism following execution is clearly another example: 100% of the participants stop committing offences, without a need for a comparison group. However, such powerful interventions are rare.

Counterfactuals

Ideally, we should follow the *same* people along two *different* parallel universes, so that we could observe them both with and without attending AA. These are true counterfactual conditions, and they have been discussed extensively elsewhere (Dawid, 2000; Lewis, 2013; Morgan & Winship, 2015; Salmon, 1994). A counterfactual is 'a comparison between what actually happened and what would have happened in the absence of the intervention' (White, 2006, p. 3). These circumstances are not feasible in real-life settings, but the best Hollywood representation of this idea is the 1998 film *Sliding Doors*, starring Gwyneth Paltrow. In that film, we see the protagonist 'make a train' and 'not make a train', and then we follow her life as she progresses in parallel (Spoiler alert: in one version, she finds out her partner is cheating on her, and in the other, she does not).[1] However, without the ability to travel through time, we are stuck: we cannot both observe the 'train' and the 'no-train' worlds at the same time.

In the context of the AA example above, we can use *another* group of people who did not take part in AA and look at their drinking patterns at the same time as those

---

[1]One of the authors uses the trailer for this film in teaching to illustrate ideal counterfactuals, with credit to Mike Hout for the idea.

who were in the AA group – for example, alcoholics in another town who did not take part in AA. This approach is stronger than a simple before–after with one group only. It is also a reasonable design: if the comparison group is 'sufficiently similar' to the AA participants, then we can learn about the effect of AA in relative terms, against a valid reference group.

The question is, however, whether the alcoholics in the other town who did not take part in AA are indeed 'sufficiently similar'. The claim of similarity between the AA participants and the non-AA participants is crucial. In ideal counterfactual settings, the two groups are identical, but in non-ideal settings, we should expect some differences between the groups, and experiments attempt to reduce this variation as much as possible. Sometimes, however, it is not possible to artificially reduce this variation. For example, our AA participants may have one attribute that the comparison group does not have: they *chose* to be in the AA group. The willingness to engage with the AA group is a fundamental difference between the AA treatment group and the comparison group – a difference that goes beyond merely attending AA. For example, those at AA might be more open to personal change and thus have a generally better prognosis to recover from their drinking problem. Alternatively, it might be that those in the AA group drank much more alcohol pre-intervention; after all, they were feeling as though they needed to seek help. It could also be that the AA group participants were much more motivated to address their drinking issues than other people – even amongst those who have the same level of drinking problems. This set of issues is bundled together under the term *selection bias* (see review in Heckman, 1990; Heckman et al., 1998). In short, the two groups were different before the AA group even started treatment, and those differences are linked to differences in their respective drinking behaviour after the treatment has been delivered (or not delivered).

In this context, D.T. Campbell and Stanley (1963) note the possibility that 'the invitation itself, rather than the therapy, causes the effect' (p. 16). Those who are invited to participate may be qualitatively different than those who are uninvited to take part in the intervention. The solution is to create experimental and control groups from amongst seekers of the treatment, and not to assign them beforehand. This 'accounts for' the motivation part of treatment seeking behaviour. To avoid issues of withholding treatment, a waitlist experiment in which everybody ultimately gets treatment, but at different times, might be used. This approach introduces new challenges, which are discussed in Chapter 4.

Still, not being able to compare like with like undermines our faith in the conclusion that AA group participation is causally linked to less drinking – but we can problematise this issue even further. In the AA example, we might have the data about drinking patterns through observations (e.g. with daily breath testing). We can collect data and measure drinking behaviour or alcohol intake, and then match the two groups based on this variable. For example, the comparison group can comprise alcoholics who have similar drinking intake to those in the AA group. However, the problem is not so much with things we can measure that differ – although that can still be difficult – the problem is with variables we *cannot* measure but that have a clear effect on the outcomes (the 'un-observables'). Statisticians are good at controlling for factors that are represented in the data but cannot create equal groups through statistical techniques when there are no data on unobservable phenomena. For example, the motivation to 'kick the drinking habit' or the willingness to get help are often unobservable or not recorded in a systematic way. If they are not recorded, then how can we compare the AA participants with the non-participants?

Finally, participants in AA expect to gain benefits from their involvement, which is different than those who did not choose to attend meetings. If those who did not participate are in a different treatment group, they should have a different set of expectations about their control conditions. Thus, motivations, commitments and decisions are relevant factors in the success of any treatment that involves people – but they are rarely measured or observed and therefore cannot be used during the matching process (Fewell et al., 2007; M. Miller et al., 2016). We base our conclusions on observed data but are resting on the untestable assumption that *un*observed differences are also balanced between the treatment and the control groups.

Given how problematic this assumption of similarity is in unobserved measures, it is worth considering the issue further. Let us say the study was conducted in a city where several members of the AA group were diagnosed with mental health problems, such as depression, and a physician prescribed the 'group' treatment for them as medical treatment, a process known as 'social prescribing' (Kimberlee, 2015; Sullivan et al., 2005). This difference alters our expectations about the results; those with depression might be starting with a worse prognosis, for example, but end with a better one. Additionally, they might also benefit more from AA because their prior mental ill health explains why their drinking might be initially higher *and* their membership of the group is a result of a physician's prescription. These conditions may not happen at all in the comparison city – but we would not know this unless we have had access to individual participants' medical records. Imagine that we went ahead with comparing the drinking behaviour of the two groups, and even assume that they have similar levels of drinking patterns before the intervention was applied in the treatment group, but not the comparison group. If we then followed up both groups of people, we can measure their consumption in the time after the intervention takes place, remembering that only one set of people has been involved in the group. Figure 2.1 shows a hypothetical comparison. We see that being in the AA group actually seems to make people worse after attending. Given that we know they had a pre-existing diagnosis of depression and were perhaps more likely to worsen over time, we could explain this relationship away. Nevertheless, imagine if we *did not* know about the depression: our conclusion would be that the group *actually made them worse* than the comparison group – when in fact this may not necessarily be true. What is more, if we were policymakers, we might cancel funding for AA groups based on this evaluation.

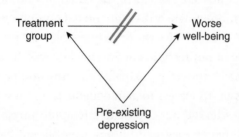

Figure 2.1   Confounding variable illustration

So what can we do? We can *randomly* assign eligible and willing participants to either AA group or no-AA group (delayed AA group in the waitlist design). This approach, which utilises the powers of science and probability theories, is detailed below.

Randomisation

The most valid method we have in science to create two groups that are as similar as possible to each other, thus creating the most optimal counterfactual conditions, is randomisation. The random assignment of treatment(s) to participants is understood as a fundamental feature of scientific experimentation in all fields of research, because randomisation creates two groups that are comparable (Chalmers, 2001; Lipsey, 1990), more than any other procedure in which participants are allocated into groups. If properly conducted, any observed differences between the groups in endpoint outcomes can be attributed to the treatment, rather than baseline features of the groups (Schulz & Grimes, 2002b). If we collect enough units and then assign them at random to the two groups, they are likely to be comparable on both known and unknown confounding factors (Altman, 1991; M.J. Campbell & Machin, 1993).

However, what properties of randomisation give credence to the aforementioned claims about the superiority of this procedure over *other* allocation procedures? The answer lies in probability theory, which predicts that randomisation in 'sufficiently large' trials will ensure that the study groups are similar in most aspects (Torgerson & Torgerson, 2003). There are different mathematical considerations to understanding probability theory (as shown in *The SAGE Quantitative Research Kit*, Volume 3). Suppose we were interested in assigning individuals into two groups, and these participants have only one characteristic: born male/female. In this example, this is binary outcome, with only two outcomes, in a distribution with approximately 50%:50% split.[2] If we allocated the pool of participants *randomly*, and with a large enough sample size, we would expect to achieve two things. First, that the overall number of participants will be fairly equal between the two study groups. Second, that in each study group, half of the participants will be male and half will be female.

Formally, this means that if we take any random variable when conducting an experiment in which the outcome is binary, researchers can rely on certain types of mathematical distributions (Bernoulli, binomial, Poisson, etc.) to calculate the probability of the outcome. When all things are equal – meaning that two groups are randomised into treatment and control conditions and *no* systematic intervention is applied to any one of the groups – then we would expect the probability of the random variable that characterises the units to 'land' in any one group to be 50%. As explained in Weisburd and Britt (2014, pp. 155–159), we can assume that as the size of the distribution increases, it will approximate a normal distribution. Put differently, in the normal distribution of a fair coin toss, the chance of success is 50% (p), while the chance of failure is equally 50% ($1 - p$) – and this is what we should expect when we conduct a random assignment: a 1:1 split between the two groups.

The benefits of random assignment have been illustrated by Shadish et al. (2002):

> When a deck of 52 playing cards is well shuffled, some players will still be dealt a better set of cards than others. This is called the luck of the draw by card players . . . in card games, we do not expect every player to receive equally good cards for each hand. All this is true of the randomised experiment. In any given experiment, observed pretest means will differ due to luck of the draw when some conditions are dealt a better set of participants than others. *But we can expect that participants will be equal over conditions in the long run over many randomised experiments* [emphasis added]. (p. 250)

For this reason (see more in *The SAGE Quantitative Research Kit*, Volume 3), Farrington and Welsh (2005) were able to conclude that 'statistical control (for baseline balance) should be unnecessary in a randomised experiment, because the randomisation should equate the conditions of all measured and unmeasured variables' (p. 30).

---

[2]We simplify for the sake of the example, however the 'natural sex' ratio at birth is approximately 105 boys per 100 girls (ranging from around 103 to 107 boys). See Ritchie (2019).

If random assignment procedures refer to the 'allocation of units according to the rules of probability theory', what does it look like in practice? These procedures ensure that each unit – or participant – has the same chance to be assigned to a treatment or control group, the same chance to *not* be assigned to one group and a *predetermined* likelihood of being assigned to any one group, just like any other participant (Allen, 2017). Therefore, think of random assignment as flipping a fair coin, or drawing a name out of a hat, and then – based on the draw – assigning the participant to one of the study groups. A more common approach uses the RANDOM function of a computer-generated software, such as Excel, www.graph-pad.com/quickcalcs/randomise1/ or www.therandomiser.co.uk, where each participant is assigned a number and then the numbers are randomly assigned to one of the treatment groups. The effect is the same: every eligible participating unit has the same chance of ending up in the treatment group or in the control group. If you do this enough times – that is, you have a considerable sample size – you unleash the magic of randomisation theory: creating two groups that are similar to one another across all dimensions – except that one group is exposed to the **stimulus**, and the other group is not.

As we noted earlier, if we assign people at random this way, then the two study groups will resemble each other: the number of people in each group would be the same (about 50% of the participants in group A and 50% of the participants in group B) – but more importantly, the *types*, *shapes* and *characteristics* of the participants would also be the same in the two groups. If the pool of participants prior to the random assignment included 30% men, 20% healthy participants and 40% college graduates, then each group should also include 30% men, 20% healthy participants and 40% college graduates (but half the absolute number of participants, thus splitting of the original pool into two groups). Then, the only difference between the groups at the post-randomisation stage is that one group is exposed to a stimulus but not the other. Crucially, the similarities are in terms of both measurable variables, such as age, gender or ethnicity, and also latent, unmeasured or unknown factors, as we suggested above when discussing the AA treatment example. With these in mind, we turn to procedures of conducting random assignment.

Procedures for random assignment (how to randomise)

There are many different procedures for allocating units into treatment and control conditions. Below, we discuss four: we have one 'pure' random assignment protocol, or the 'simple random assignment' procedure, in which units are assigned purely by chance into the treatment and control groups. As the number of units involved in testing any particular intervention grows, pure random assignment will increasingly generate the most comparable treatment and control conditions. However, researchers are aware that any *one* experiment may not necessarily create two balanced groups, especially when the sample size is not large 'enough'. For example, in a sample of of 152 domestic violence cases randomly assigned by the court to one of two types of mandated treatments between September 2005 and March 2007, pure random assignment produced 82 versus 70 participants (Mills et al., 2012), or a 1:1.17 ratio.

Therefore, as we discuss more fully in this section, it is not always possible to rely purely on chance, and strategies have been developed over the years as alternatives to pure random assignment. These are the 'restricted random assignment' protocols, and while they are common in medicine, they are far less common in experimental disciplines in the social sciences. According to these procedures, the researcher arranges the data prior to random assignment in certain ways that increase the precision of the analysis. For example, researchers often segregate the pool of participants into subgroups, or blocks, based on an important criterion; this criterion is used because there is reason to believe that some participants react 'better' to the intervention. Random assignment is conducted *after* the subgrouping has occurred, so the test is still fair and does not disadvantage any one group of participants over the other. There are other procedures like the 'randomised block design' described above: trickle flow random assignment and batch random assignment, matched pairs design and minimisation. We delve into these below.

Simple random assignment

The simple random assignment procedure is the most prevalent method of random assignment in criminal justice research (Ariel, 2009; Ariel & Farrington, 2014). In simple random assignment, the allocation of all units into treatment and control conditions is made by chance alone (Rosenberger & Lachin, 2002, p. 154; Schulz & Grimes, 2002a). Due to randomisation, every unit in the research population has the exact same chance of being assigned into the treatment or control groups, without any regard to the characteristics of the units (see Chalmers, 2001; Chow & Liu, 2004, pp. 134–135; Lachin, 1988a). With a sufficiently large sample size, this procedure will result in study groups that are roughly the same size, and these pre-experimental variables would be spread evenly between the study groups, as we discussed above.

Like all randomisation procedures, one feature of simple random assignment is the unpredictability of the allocation sequence (i.e. it is impossible to guess, prior to the allocation, whether one will fall into treatment or control conditions). This is important because in many instances people will try to cheat the randomisation process and allocate treatments based on their preferences, rather than rely solely on the random sequence. In practice, treatment providers should not be able to know how the next patient or participant will be assigned – and this benefit is crucial in order to reduce the likelihood of **selection bias**.

To consider the damaging effect of selection bias, imagine a situation where a doctor who is taking part in an experiment knows that the next patient will be offered the treatment but does not think the patient is the right 'fit' – this might lead to the doctor not offering treatment (overriding the allocation). This can happen with police experiments as well, when police officers or their partner agencies in field experiments allocate certain offenders to prevention programmes that they (genuinely) believe they would benefit from, and not based on the predetermined random assignment sequence. Interestingly, in both cases, the practitioners are hurting the very patients or offenders they are trying to help, because such 'contaminated' trials could lead to nil differences between the treatment and the control groups (as patients of both groups are receiving the same intervention). The experimenter (who is unaware of the contamination) would then conclude that the intervention did not 'work' and therefore not recommend its adoption as policy.

Case Study 2.1

Case Study of a Simple Randomisation Design: Project CARA – Conditional Cautions for Domestic Violence Perpetrators in Hampshire, UK (Strang et al, 2017)

Faced with a growing number of domestic violence cases, most of which resulted in no 'official' detection, the Hampshire Constabulary set out to design and test a workshop-based programme for domestic abusers. Working alongside The Hampton Trust charity and the University of Cambridge, officers from Hampshire Constabulary tested the issuance of conditional cautions to low-risk offenders. Conditional cautions are a version of criminal cautions that are not progressed further to prosecution if specified conditions are met by the offender. The Hampton Trust developed a 2-day workshop raising the abusers' awareness of their own behaviours, and the university assisted with the design of the experiment (Strang et al., 2017).

In this experiment, a computerised algorithm was used to assign eligible offenders to either treatment or control groups, using the **simple random assignment** technique. Both groups were issued with a caution not to commit any further offences within four months; however, the treatment group participants were *also* required to sign a statement of guilt and attend the workshops. Control group members were not required to attend a workshop. In total, 4768 cases were screened for eligibility, of which 1469 were deemed eligible, and 293 were referred for randomisation during the experimental period. In a 1-year follow up after the random assignment, the treatment group caused less harm to victims than the control group. These results have led to the exploration of similar schemes by other police forces in England and Wales.

Alternatives to simple random assignment: restrictive allocation

As useful as simple random allocation is, it does not always create balanced groups, especially with smaller samples. When the data are characterised by high variability – that is, participants with different backgrounds and large differences between them – it decreases the ability of the researcher to create balanced groups using simple random allocation (Lachin, 1988a). This means that in one of the groups, we may end up with more men, younger participants, hotter crime hot-spots or more chronic offenders, for example. This is a direct function of the sample size: the smaller the study, the more likely that simple random assignment will allocate more extreme scores to one group than to the others.[3] To resolve this, and to capitalise on the heterogeneity that exists in the data, alternatives to the simple random allocation procedure were developed (see Garner, 1990). These are referred to as 'restricted allocation procedures'. The many different types of restricted randomisation methods include block randomised trials (see Ariel & Farrington, 2014), stratification (J.L. Gill, 1984; Green & Byar, 1978; Lachin & Bautista, 1995) and minimisation (Pocock & Simon, 1975; Scott et al., 2002). We discuss these and others below.

More formally, restricted allocation procedures work to correct the limitations of simple random assignment (see Lachin, 1988b; Lipsey, 2002; Meinert & Tonascia, 1986). In these procedures, the researcher still uses randomisation, but in a more structured way (Friedman et al., 1985) – hence the name: a 'restriction' on the pureness of the random allocation of all cases in the pool. If the researcher knows *ahead of time* that a particular variable can affect the outcome of the experiment, they can therefore separate the participants according to this criterion, and *then* conduct the random assignment. For instance, if the researcher suspects that criminals of different backgrounds (e.g. property offenders vs violent offenders) will be affected differently by the intervention, then they could first separate the overall sample according to this criterion (offence type), and then conduct random assignment within each category. This procedure, in turn, will reduce the 'noise' in the original sample that exists due to having different offender types grouped together. This procedure can therefore create two separate experiments running at the same time – one for property offenders and one for violent offenders. The experiments have an identical experimental protocol, process and treatment – except a more

---

[3]You can recreate this issue with a coin toss. If you flip a coin 10 times, you might end up with 9 heads and 1 tail, or 3 of one and 7 of the other. Now, imagine the 9:1 allocation represents a characteristic like 'being a frequent offender' – we end up with 9 in treatment and 1 in control – which we would expect to have quite a large impact on our results.

structured approach is taken in the random allocation of the cases, given the importance of this criterion.

As we noted, imbalances are more likely to occur with a simple randomisation procedure in small sample trials (Farrington & Welsh, 2005; Lachin et al., 1988). Therefore, when there are 'several hundred participants' or fewer, restricted randomisation techniques should be preferred over simple random allocation (Friedman et al., 1985, p. 75; R.B. Santos & Santos, 2016). The question of sample size depends on the context, the units of randomisation, and the putative effect size (as shown in a series of studies by David Weisburd and his colleagues; see C.E. Gill & Weisburd, 2013; Hinkle et al., 2013; Weisburd, 2000; Weisburd & Gill, 2014).

The randomised block design

The randomised block design is one of the most frequently used experimental designs, at least in biomedical research (Abou-El-Fotouh, 1976; Cochran & Cox, 1957; Lagakos & Pocock, 1984), but it remains rare in criminal justice research (Ariel, 2009; Weisburd et al., 2015). A famous example also used by Neyman (1923), Fisher (1935) and Hill (1951) can best describe the process, which is taken from the world of agriculture (Ostle & Malone, 2000):

> Six varieties of oats are to be compared with reference to their yields, and 30 experimental plots are available for experimentation. However, evidence is on file that indicates a fertility trend running from north to south, the northernmost plots of ground being most fertile. Thus, it seems reasonable to group the plots into five blocks of six plots each so that one block contains the most fertile plots, the next block contains the next most fertile group of plots and so on down to the fifth (southernmost) block, which contains the least fertile plots. The six varieties would then be assigned at random to the plots within each block, a new randomisation being made in each block. (p. 372)

Thus, like simple random assignments, where units are unrestrictedly distributed at random to either treatment or control (or more than two groups, as the case may be) group, under the randomised block design, units are still allocated randomly to either treatment or control group. However, this allocation is conducted within pre-identified blocks. The blocking process is established and based on a certain qualitative criterion, which is intended to separate the sample, prior to assignment, into subgroups that are more homogeneous than the group as a whole (Hallstrom & Davis, 1988; Simon, 1979). Blocking is therefore a design feature that reduces variance of treatment comparisons by capitalising on the existing blocks (or strata), which are less heterogeneous than the entire pool of units.

Box 2.2

Stratification Versus Blocking

There seems to be some confusion between two similar but not entirely identical processes: 'stratification' and 'blocking'. While both processes create strata or blocks of participants, and therefore increase the control over the heterogeneity in the data, *stratifying* is often a term reserved to sampling of units from a given population, as a way of increasing the representativeness of certain variables, whereas the allocation of cases in experiments is commonly referred to as 'blocking'. In practice, the words "strata" and "blocks" [and stratification and blocking] are used synonymously and have the same benefit - creating more balanced treatment and control groups at the point of randomisation.

We also note that there are complicated versions of blocking, including 'Latin squares' and 'permuted block random designs'. These are more fitting when there are two or more treatments administered simultaneously, and when the scholar is interested in measuring their independence as well as their **interaction effects** on the dependent variable. For further reading, see Ariel and Farrington (2014) and Chapter 4.

Equally important, everything else in the experiment remains constant. This means that the treatment (its dosage, delivery frequency or method of administration), the instruments used to measure the pre-treatment and the post-treatment observations (e.g. consistent measure by the same observation instrument) and the management of the study in continuous and uninterrupted ways remains identical throughout the experiment. When it comes to block random assignment, this constancy remains intact, but within blocks of data (Matts & Lachin, 1988; Ostle & Malone, 2000; Rosenberger & Lachin, 2002).

Note that blocking can also allow the researcher to oversample a particular subgroup that is relatively small in the overall population (e.g. uncommon ethnicities, prognoses or extreme crime values). Blocking can therefore allow for a clear representation of different levels of variables and then a fair comparison between these different levels. Further note that within the blocks there is usually only one treatment group and one control group because the test statistic is considered more 'stable' this way (see Gacula, 2005; however, cf. Grommon et al., 2012). There are compilations of multiple studies, which D.T. Campbell and Stanley (1963) refer to as factorial designs. These designs are not discussed in this book.

Imagine a trial on the effect of a domestic violence counselling treatment for male offenders on subsequent domestic violence arrests. Such an experiment would benefit from blocking the sample before random assignment to the study groups, according to number of prior domestic violence arrests. There may be a clear reason to assume that persistent batterers will respond differently to counselling (likely to be less susceptible to treatment). One approach can be to block the sample based on their prior arrest history: for example, 'substantial history', 'some history' and 'no additional history' of previous domestic violence[4] and then randomly assigning treatment and no-treatment within each block. We can also further block the data according to a second criterion (e.g. whether the abuser has already undergone some sort of treatment for domestic violence in the past); however, ordinarily, there will be a single blocking factor (Ostle & Malone, 2000). With the blocking procedure, the intra-block variance of pre-intervention domestic violence levels (measured in terms of prior domestic violence arrests) is expected to be lower than the variance across the entire sample (Canavos & Koutrouvelis, 2008) – that is, the blocking creates more homogeneous blocks of domestic violence levels and allows the researcher to compare the intervention at these different levels. Another way to describe this is by saying that the 'signal' (i.e. the intervention effect) was not changed; however, the 'noise' (i.e. the variance)[5] is decreased. The researcher is able to gain more precision in their estimate of the treatment effect.

Case Study 2.2

Case Study of a Randomised Block Design: Problem-Oriented Policing in Violent Crime Places (Braga et al., 1999)

In 1999, Professor Anthony Braga and colleagues set out to evaluate the impact of problem-oriented policing (POP) responses on violent street crime (see a recent review of POP in Hinkle et al., 2020; see POP in place-based context in Braga, Turchan, et al., 2019). In the 1990s, POP strategies had been found effective to reduce general crime and disorder (Eck & Weisburd, 1995), but it had not been tested specifically in the context of violence. Working with the Jersey City Police Department, the researchers developed a blocked RCT. Violent crime hotspots (defined as clusters of individual street intersections) were mapped using computerised crime mapping processes, resulting in a list of 268 intersections in 56 discrete hotspots (these locations accounted for 6% of the intersections in Jersey City but 24% of assaults in Jersey City). Jersey City police officers were then invited to identify the problems at each location and then 'diagnose' their causes.

To control for variation among the 56 hotspots, the researchers used a randomised block design. The hotspots were blocked based on a range of qualitative factors such as knowledge of the types of robbery taking place, the dynamics of the activity and the physical makeup of the location. At the treatment locations, officers were instructed to develop and execute crime prevention plans. The experiment observed statistically significant reductions in street fights, property and robbery crimes, with variations explained by the differing nature of the situational responses implemented by the Jersey City officers.

---

[4]In this example, the arrest history is a continuous count variable and the three 'blocks' are assigned mutually exclusive qualitative values – much like the fertility rates of the field segments in the original Fisher (1935) and Hill (1951) illustrations of randomised block designs in agriculture.

[5]In statistical terms, *noise* normally refers to unexplained variation in data samples.

Case Study 2.3

Case Study of a Partial Block Design: Acoustic Gunshot Detection Systems in Philadelphia, USA (Ratcliffe et al., 2019)

Seeking to assess the impact of acoustic gunshot detection systems on police department workloads, Professor Jerry Ratcliffe and colleagues set out to randomly assign 20 sensors between 40 CCTV (closed-circuit television) camera sites. The sites were matched using clustering based on counts of gun crime, socio-economic status and the proportion of land used for residential purposes. However, operational assessments undertaken after the (blinded) random assignment had taken place concluded that some sites were unsuitable for sensor deployment. Five sites required the swapping of control and treatment status due to technical reasons, and in two sites, miscommunication resulted in sensor deployment in both treatment and control areas (these were later matched to new control sites). The researchers referred to this model as a 'partial block design'. The findings concluded no significant difference in demand levels for the Philadelphia Police Department, but it is difficult to draw concrete conclusions from a broken experiment (see also Barnard et al., 1998). This study serves as an example of the practical difficulties researchers face in designing experiments for real-life settings. In this context, partial blocking can allow research designs to be 'saved' when logistical issues obstruct the original designs.

Trickle flow random assignment

Another common research design is the trickle flow, where random assignment is conducted as eligible cases come into the experimental pipeline sequentially. This randomisation procedure is useful when units are not ready to be assigned all at once, For example, a study on the effect of rehabilitation programmes delivered after release from prison would not have enough prisoners released on any given day to allocate them into a sufficiently large study through simple random assignment. Similarly, an experiment on the effect of advice given to victims of burglary on how to reduce the likelihood of repeat victimisation would struggle to fill all the experimental slots in one go (thankfully, there are not enough eligible burglary victims on any given day!).

Another feature of this sequential design is the ability to cut short recruitment when the observed differences between the experimental and control conditions are sufficiently large that the experiment should be stopped. There are ethical merits to this feature. We would like experiments in which the treatment effect is so potent (whether beneficial or counterproductive to the participants) that the study would have to stop. In practice, this is unlikely in the social sciences. The effect sizes are rarely that large, meaning that the magnitude of the differences between treatment and control conditions are not as robust as in medicine. There are also strong methodological counterarguments that call for experiments to continue until all experimental slots are filled. First, treatment effects in field experiments are usually not 'large' (e.g. Weisburd et al., 2016), so the dilemma about curtailing the experiment before the planned randomisation sequence has been fully filled is almost unimaginable.[6] We visit these ethical considerations in Chapter 5.

Second, most field experiments take a minimalist approach: recruit the least number of units needed for the trial to result in statistically significant differences between the two groups (should such a difference exist). Large trials are typically more expensive and more difficult to manage (Weisburd et al., 2003), so researchers carefully plan for 'just the right size'. Therefore, early indications about large and observable differences between the treatment and control groups, but which are based on small samples, can be highly suspect. Extreme results can simply be a natural extreme variation in the data. However, this issue will naturally resolve when more cases are recruited into the study. For example, in an experiment on the effect of police body-worn cameras on assaults against police recruits, several new officers from the same treatment group may be attacked immediately after being recruited into the study; this scenario will inflate the mean of the entire group because of the limited time of observation (i.e. assaults against recruits are infrequent, and they are less likely as the study period is shortened). By chance alone, an unusual number of units from the same study group exhibited extreme scores, but this extreme overall difference in scores is likely to 'flatten out' when more units are assigned into the two groups over time. This issue, which is largely a **statistical power** consideration, is visited in Chapter 3.

The trickle flow can take two major forms: case by case or in sequential batches of cases. When eligible units enter the study one at a time, then each allocation is done randomly at the moment the unit becomes eligible and available for recruitment. For example, as soon as an offender is arrested, they are then randomly allocated into treatment or no-treatment conditions (as in the case of the CARA experiment depicted in Case Study 2.1; see also Braucht & Reichardt, 1993; Efron, 1971).

Trickle flow in batches means that the experimenter conducts the random assignment of multiple cohorts that enter the study. An example is a training programme for new police officers, which is primarily delivered to many candidates simultaneously (see Case Study 2.4): as the police department trains new recruits in multiple batches throughout the year, each incoming batch may be slightly different than the other batches, so random assignment is conducted within these cohorts. This design and its analysis are similar to the block random assignment procedure, but the effect of the temporal sequencing of the batches should be taken into account as well.

Finally, notice that in trickle flow studies there is a special need to create balance between the treatment and the control groups in terms of the 'time-heterogeneity'

---

[6]Although we note that with the use of administrative data it is now possible to run experiments with hundreds of thousands, or even millions, of participants.

bias, which does not occur when units are allocated in one 'go'. This bias is created by changes that occur in participants' characteristics and responses between the times of entry into the trial (Rosenberger & Lachin, 2002, pp. 41–45). Therefore, we need to have balance not just in terms of the total number at the end of the experiment but throughout the recruitment process as well (Hill, 1951). This issue was explored in various statistical notes (Abou-El-Fotouh, 1976; Cochran & Cox, 1957; Lagakos & Pocock, 1984; Matts & Lachin, 1988; Rosenberger & Lachin, 2002, p. 154). In practical terms, however, in order to avoid the time-heterogeneity bias, randomisation can be restricted in a way that requires that every so often (e.g. every 20 units) the split between treatment and control conditions will be 50%:50%.

Case Study 2.4

Case Study of a Batch Trickle Flow Design: Diversity Training for Police Recruits (Platz, 2016)

Seeking to address concerns that stemmed from the public debate about bias, diversity and policing behaviour, the Australian Queensland Police Department conducted an RCT of a 'values education programme' given to new police recruits. A group of 260 new police officers were assigned to treatment (in which they participated in a 2-week diversity course during their initial training) or control group (in which they did not complete the course during training). The recruit cohorts entered the programme in three batches and thus were randomised congruent with their intake dates. Results from the 132 experimental and 128 control participants overall indicated that support for diversity declined over the duration of the initial training; however, it was less so for those individuals receiving the diversity course.

Matched pairs design

Another design commonly used in criminal justice experiments is the randomised pair, or matched pairs design (e.g. Bennett et al., 2017; Braga & Bond, 2008; MacQueen & Bradford, 2015; Telep et al., 2014),[7] which is a special type of randomised block design. It can be used when the experiment has only one treatment condition and one control condition, and participants can be grouped into pairs based on the levels of the dependent variable at its pre-test level. The participants are first *rank ordered* according to their values on this variable, and then, within each pair, participants are randomly assigned to different treatments (Conover, 1999). Put differently, the participants are listed in descending order according to the variable

---

[7]On the history of matched paired designs see Welsh et al. (2020).

of interest, and then every multiple of two units are paired (e.g. units 1 and 2, 3 and 4, 5 and 6 and so on), with each pair forming a 'block' (pair A is 1 and 2, pair B is 3 and 4 etc.). Within each pair, units are placed into treatment or control. This process ensures balance on this single continuous variable. (It is worth noting that the major drawback of this design is that if one unit from the pair retires from the study then *both* units are lost to analysis, unless statistical corrections for missing data are used *ex post facto* – an approach that controlled experiments aim to avoid.)

An example from experiments on the effect of hotspots policing illustrates the utility of this approach (and its perils, too). Ten pairs of hotspots (i.e. 20 hotspots) are ranked according to the number of crimes that took place in the hotspots pretest, and then randomly assigned in order to test the effect of saturated police presence in treatment hotspots compared to control hotspots, which receive ordinary policing tactics. The outcome of interest is the number of crimes reported to the police from the hotspots. For each pair of hotspots (which were, again, rank ordered in terms of the 'heat' at the dependent variable's pretest scores), one is randomly allocated to the treatment group and then visited by an officer several times during the day, for 15 minutes per visit. After three months, all 20 hotspots are analysed, as shown in Table 2.1.

Table 2.1   Illustration of a small-sample hotspots policing experiment using paired random assignment

Pair Number	Control Hotspots Crime Count	Treatment Hotspots Crime Count	Difference
1	98	84	14
2	84	74	10
3	73	80	−7
4	80	65	15
5	79	67	12
6	74	63	11
7	61	41	20
8	52	48	4
9	56	43	13
10	43	33	10
Total crimes	700	598	102
Mean crimes per hotspot	70.0	59.8	10.2

The benefit of this approach is clear: the noise is dramatically reduced – which can be seen in the range of scores. Given the range of hotspots in this city in terms of the number of crimes these areas are exposed to, the sample can be said to be heterogeneous. Therefore, within pairs that are more closely similar in terms of the outcome variable, the crime levels are much more similar than the overall group. As shown, this pairwise random assignment is particularly useful for situations where the focus is reducing statistical variability in the data. We increase the statistical power of the test, as the intra-block variance – that is, between the two pairs – is lower than the simple randomisation technique. Experimenters can therefore benefit from this approach – especially in small n studies and when there is a clear dependence between the two pairs (twin studies, co-offenders, co-dependent victims etc.). We return to the issue of statistical power in Chapter 5.

This benefit, however, comes with a cost, which must be taken into consideration when considering this design. When the two units are paired based on a particular blocking criterion, they should subsequently be viewed as matching based on that criterion *only*, not any other variable. This means that the experimenter will not be able to claim balance based on any other variable. Take, for example, a recent RCT on the effect of hotspots policing in Sacramento, California (Telep et al., 2014). The study was designed to reduce crime in the city during a 90-day RCT, using saturated police presence at hotspots by spending about 15 minutes in each hotspot and moving from hotspot to hotspot in an unpredictable order to increase the perception of the costs of offending in those areas (approximately one to six hotspots in their patrols, once every Two hours). Officers were not given specific instructions on what to do in each hotspot; they received daily recommendations through their on-board computers to engage in proactive activities such as proactive stops and citizen contacts.

A total of 42 hotspots were identified as eligible units (being the 'hottest' hotspots in the area), with half assigned to treatment and half to business-as-usual conditions. In order to reduce the variability between the 21 treatment hotspots and the 21 control hotspots, the sites were paired prior to randomisation based upon similarity in levels of calls for service, crime incidents and similar physical appearance based on the initial observations. After pairing, a computerised random-number generator assigned hotspots to either the treatment or the control group. The pairing 'worked': dependent or paired samples t-tests showed no statistically significant differences between the treatment and the comparison hotspots in calls for service or serious crimes in 2008, 2009 or 2010, suggesting no reason for concern about pre-randomisation baseline differences between the group (although see section below on significance tests for baseline imbalances after randomisation). The results of the intervention and tightly managed experiment suggested significant overall declines in both calls for service and crime incidents in the treatment hotspots relative to the controls.

However, the matching criteria did not create similar hotspots in terms of other important features – for example, time spent at hotspots, the type of activities that officers delivered at the hotspots or the frequency of visits that officers made to the hotspots. In fact, there were stark baseline differences (Mitchell, 2017). While we may assume that treatment and control hotspots should have received, on average, similar levels of the attention (measured by the time spent at the hotspots), in reality, this was not the case. This is important, because if we are now interested in measuring other outcomes – for example, a link between post-randomisation dosage in terms of the GPS-recorded time spent at hotspots and differences in outcome variations – then we are not able to do that, unless we implement statistical controls for the baseline inequality that the pairing has inherently created in the data. Thus, this was a flaw in the design: not considering ahead of time the likelihood of analysing other outcome variables. If these were measurable, then they could and should have been incorporated into the study design using, for example, minimisation or stratification. These controls become challenging with a sample size of 42, especially since the whole premise of RCTs is *not* to implement statistical controls in order to create equilibrium at baseline.

Case Study 2.5

Case Study of a Matched-Pair RCT Design: Procedural Justice Training (Antrobus et al., 2019)

Police in Queensland, Australia, set out to field test the effect of procedural justice training for new recruits. Procedural justice techniques had long been established as having positive effects on legitimacy and associated compliance with norms (Hough et al., 2016; Tyler et al., 2014). However, only a handful of other studies had examined the effect of specific training programmes, and gaps still remain as to the effects of such programmes in real-world scenarios. The Queensland Police Department sought to devise a programme to impact 'business as usual' practices and so developed lessons and materials as a core part of their new recruit training programme.

The randomised experiment aimed to evaluate the impact of the programme first, on the procedural justice displayed in the interactions between those new recruits trained in it and the public, and second, on the recruits' attitudes. However, the sample size available was small ($n = 56$), so the researchers sought to maximise statistical power by using a matched-pairs design. The 56 new recruits were matched based on prior knowledge of equivalence **covariates** such as gender, academic record and age. A prospective analysis of statistical power suggested that 26 participants in each group would be sufficiently powerful to detect a significant effect size. In the final experiment, 28 pairs were achieved with no statistically significant difference in the covariates found between the treatment and the control groups (notwithstanding that the small sample size means that it may be difficult to detect valid differences). Using the recruits' mentors to rate interactions, the experiment found that recruits exposed to the treatment were more procedurally just. However, there was no significant differential effect on officer attitudes.

Minimisation

Introduced by Pocock and Simon (1975) and Taves (1974), minimisation is a 'dynamic' randomisation approach that is suitable when the experimenter wishes to stratify many variables at once within a trickle flow experiment. In this scenario, the process of minimisation is as follows. The first case is allocated truly randomly. The second case is then allocated to both treatment and control, with whichever allocation minimises the differences between the putative treatment and the control groups (hence *minimisation*) being favoured. The third case is allocated in the same way and so on. If at any point the groups are equal on the balance measure – which is a summary of the standardised differences in means between groups – the next allocation is again truly random.

To prevent this process from being deterministic (and thus open to cheating), there is a degree of unpredictability built in. That is, even when it seems obvious that a case would be allocated to control, it might still be allocated to treatment. (This is sometimes governed by allocation favouring treatment over control using a 'biased coin'; see Taves, 2010.)

Case Study 2.6

Case Study of a Minimisation RCT Design

In 2011, a team of researchers tested a new intervention to reduce school exclusion, an approach used in the UK for more than 200 years with no evidence of effectiveness. The project, titled the London Education and Inclusion Project, given the involvement of Greater London Authority, Catch22 charity and scholars from the University of Cambridge (see Obsuth et al., 2016), was an RCT of around 600 pupils at high risk of exclusion in 36 schools. Researchers found that in the short term, the intervention was ineffective and may have had harmful effects for some subgroups (using self-reported data; see Obsuth, Sutherland, et al., 2017). In the long term, there was also no detectable difference for outcomes such as school attainment.

The design challenge was that there were several variables related with school exclusion on which the researchers had to create balance: percent of pupils eligible for free school meals, the percent of pupils with special educational needs, whether the school was mixed or single sex, the size of the school and levels of antisocial behaviour (Obsuth et al., 2016). At the point of randomisation, schools were randomised via minimisation using MinimPy (an open-source minimisation programme in Python programming language) and the procedure described above. As schools were the unit of randomisation, balance was assessed at the school level (Table 2.2).

(Continued)

Table 2.2 School-level balance in the London Education and Inclusion Project trial

| | Balance Factors | | | | | | | | | | |
| | FSM | | School Sex | | ASB (PCA) | | SEN | | School Size | | | |
Allocation	<37 %	≥37 %	Mixed Sex	Single Sex	<Mean ASB	≥Mean ASB	<12.5 %	≥12.05 %	Large	Medium	Small	Total
Intensive	8	9	12	5	10	7	9	8	8	6	3	17
Light	9	10	14	5	9	10	9	10	7	7	5	19
Total	17	19	26	10	19	17	18	18	15	13	8	36

Note. FSM = free school meals; ASB = antisocial behaviour; PCA = Principal Components Analysis; SEN = special educational needs.

Randomisation in practice: the 'randomiser'

To this point, we have discussed what randomisation is and why and how it works; however, we have not given much exposure to the issue of practicality (although the choice of the units of randomisation is very much realpolitik). Tossing coins or drawing names from hats or sealed envelopes is not common practice. Online solutions like the 'randomiser' (Ariel et al., 2012; Linton & Ariel, 2020), an open-source software that allows treatment providers to conduct random allocation processes themselves, can provide a solution for most randomisation procedures.[8] Given the substantial costs associated with random assignments using humans, randomisers are user-friendly, safe and cheap platforms that enable researchers and their partners to conduct the allocation themselves. The integrity of the random allocation procedure can be preserved, as the research team maintains full control over the process at the back end.

Units of randomisation (what to randomise?)

Once we acknowledge that experimenters should pay close attention to the ways in which they conduct random assignment, the next question is 'What can they randomise?'

A fundamental aspect of the design of all experimental methods is a:

> "clear identification of the experimental unit. By definition, this is the smallest object or material that can be randomly and independently assigned to a particular treatment or intervention in the experiment." (N.R. Parsons et al., 2018, p. 7)

---

[8]For example, www.therandomiser.co.uk, www.randomiser.org/ or www.random.org/lists/

In criminology, it is often the case that individuals – officers, victims, patients or members of the general public – are considered the 'units', which are then entered into the experiment (for a review of units used in criminology, see Weisburd, 2015). However, individuals as the units of analysis are only one option and, often, not the best choice given the various challenges of this approach. There are also theoretical reasons for choosing different units – entire groups, places, classrooms, dyads or other types of units. Here, we explore three of these units – people, places and time – as an illustration of the dilemmas that experimenters face when choosing the optimal unit of analysis in their research. We use a hypothetical study on the effect of the use of Tasers in policing to problematise these concerns.

Researchers can hypothesise that the possession of Tasers reduces aggression in volatile police–public engagements (see Ariel, Lawes, et al., 2019). When the less-than-lethal weapon is deployed in field settings, officers are in a position to effectively subdue uncooperative suspects with a high degree of success – which leads to the assumption that suspects are deterred from assaulting the officer or resisting lawful orders, as a result of the presence of the Taser. Yet how would we go about testing these effects in real-life settings? How would researchers, for example, be able to single out the effect of the Taser, ensuring that no confounding variables affect the results? The researchers can choose between individual officers, crime hotspots or time shifts to randomly allocate these units into 'Taser' and 'no-Taser' groups.

Person-based randomisation

Intuitively, the simplest approach would be to randomly allocate individual police officers into treatment and control groups. This way, whatever n of eligible front-line officers the participating department has, the researchers would then allocate 50% of n to a Taser group and the remaining 50% of n to the no-Taser group. The use of force by these officers will then be measured during a reasonable follow-up period, and any variations in the rate of force used by officers in police–public contacts would then be attributed to the treatment (Tasers), given the random allocation (Henstock & Ariel, 2017).

However, this research design is not so simple in practice. On paper, indeed the randomisation of individual officers is ideal because their allocation at random should cancel out selection biases, differences between the officers themselves or other factors that may affect the use of force in police–public encounters. However, one issue that cannot be ignored is the interference, or the crossover, between treatment and control officers. While in many police departments the officers work in solo formations (i.e. a single-officer patrol unit), many departments deploy officers in double-officer formations (Wain & Ariel, 2014). Therefore, individual officers cannot be the unit of analysis in these departments when the basic unit of patrol is a double-crew patrol. If an officer in the treatment group is paired with an officer in the non-treatment group (as they normally would), there will be a contamination effect. When this patrol *unit* attends a call for service or conducts a stop and frisk, it is as if both officers are operating under the treatment conditions. Thus, a simple random assignment procedure would have to be a patrolling *unit*, rather than the individual *officer*. Otherwise, there could be a scenario where one officer was randomly assigned into treatment conditions (Taser), while their partner was randomly assigned into control conditions (no-Taser).

This contamination is detrimental to the study. We hold the view that any experiment that ignores this issue is doomed to failure – and it is difficult to correct for the spillover once the experiment has been completed. In practical terms, under these conditions, the scholar has lost control over the administration of the treatment; they cannot reliably tell the difference between treatment and control conditions.

To complicate this further, consider additional risks to the independence of the treatment and control conditions in both models (single- or double-crew formations). Operational needs within emergency response police systems often require ad hoc, triple crewing or even larger formations, particularly when responding to complicated crimes. This means that officers in the control group could have been 'contaminated' by responding to calls together with any number of members of the treatment group – and vice versa (Ariel, Sutherland, & Sherman, 2019). As the treatment is hypothesised to affect interactions with members of the public, 'control officers' would have had their behaviours altered in response to the presence of their colleagues' Tasers. At the very least, suspects and victims would behave differently when Tasers were present (Ariel, Lawes, et al., 2019), even if only some officers were equipped with them. The medical analogy is a clinical trial where both the treatment and the control patients are sharing the same pill, even though it was assigned to treatment patients only.

Finally, another reason why using the individual officer as the unit of analysis seems illogical is that it dismisses group dynamics and organisational factors that are very difficult to control for (Forsyth, 2018). These may include the character of the officer or the sergeant managing the shift, the degree of the officers' cynicism, comradery, codes of silence and a host of institutional undercurrents that are recognised in the literature (Ariel, 2016; Skolnick, 2002; Tankebe & Ariel, 2016). There are underlying forces and cultural codes of behaviour that characterise entire shifts, and as most of these factors are not recorded, they therefore cannot be included in statistical models that aim to control for their confounding effects (see Maskaly et al., 2017). Thus, in many studies, using the officer as the unit of randomisation highlights a misunderstanding of the ways in which police officers are deployed in the field. Police are typically deployed in formations that, a priori, create a weak research design for individually randomised trials, leading to poor intervention fidelity, crossovers and spillovers (see more broadly Bellg et al., 2004; Shadish et al., 2002, pp. 450–452).

Place-based randomisation

Another unit of analysis that can be utilised in a study of the effect of Tasers is location. Criminological research has shown that crime is heavily concentrated in discrete areas called hotspots (Braga & Weisburd, 2010; Sherman et al., 1989; Sherman et al., 1995; Weisburd, 2015). For example, Sherman et al. (1989) are credited with being the first to systematically explore these concentrations, identifying that half of all calls for service come from less than 3.5% of addresses in a given city. Weisburd et al. (2004) subsequently demonstrated that 50% of the crime recorded in Seattle, USA, occurs at 4.5% of the city's street segments (see also Weisburd et al., 2012). Hotspots have also been shown to be relatively stable over time and location (W,eisburd et al., 2012). Notably, Weisburd et al. (2004) also found that Seattle street segments that recorded the highest amount of criminal activity at the beginning of the authors' longitudinal study were similarly ranked at the end of it. Such micro-places remain stable because they provide opportunities for criminal activity that other areas may lack (Brantingham & Brantingham, 1999). Thus, a large body of research on hotspots clearly demonstrates that crime and disorder tend to concentrate in very small, predictable and stable places. Such crimes include violence, robbery and shootings (see Braga et al., 2008; Rosenfeld et al., 2014; Sherman et al., 2014).

Given this line of research, a focus on hotspots 'provides a more stable target for police activities, has a stronger evidence base and raises fewer ethical and legal problems' (Weisburd, 2008, p. 2). Evidence shows that when experimenters nominate hotspots as the unit of analysis and then apply directed police interventions, this can result in effective crime reductions (Braga, Weisburd, & et al., 2019). For example, in the Philadelphia Foot Patrol Experiment (Ratcliffe et al., 2011), violent crime fell by 23% in the treatment area after three months of dedicated patrolling relative to the comparison areas. An RCT of directed police presence at the 'hottest' 115 platforms within the London Underground was also shown to cause a significant overall reduction in crime and calls for service (Ariel, Sherman, et al., 2020; Braga, Turchand, et al., 2019).

In terms of research methods, a study on the effect of Tasers can start by concentrating on violent hotspots, then subsequently allocating officers equipped with Tasers to half of the identified locations. It is important that the hotspots are allocated accordingly (as discussed above), which means that officers with Tasers will go to treatment hotspots and officers without Tasers will go to control hotspots. We can then measure the utility of the deployment of officers with Tasers to reduce the use of force in the hotspots, compared to hotspots without this intervention.

Case Study 2.7

Hotspot Policing: The Next Generation Focusing on Harm in Experimental Criminology

We argue that a major shift of emphasis is taking place in the application of hotspot policing. Police forces continue to measure themselves against and focus on the counting of crime incidents as a measure of success or failure. Each year, crime reductions are discussed alongside the overall number of crimes. Likewise, particular areas of work are targeted based on these counts: more events lead to a higher number of police interventions. However, we argue that focusing merely on counts, rather than on the severity or harm of crime, is somewhat crude and imprecise. It also makes policing more expensive than it should be. As posited recently by Sherman (2013), 'all crimes are not created equal[;] some crimes cause horrible injuries and deaths. Others cause scant harm to anyone' (p. 422). In an age when resources are scarce, not all crimes can attract (or deserve) the same reaction. As such, a triage approach is required whereby the most harmful events are treated first, including crimes against persons such as violence, robbery and weapon crimes (on different triaging scales in emergency medicine, see Farrohknia et al., 2011).

Given this new approach, we denote the 'classic' way of defining places with higher crime count concentrations than others as 'hotspots' and classify places characterised by higher crime harm concentrations as 'harmspots'. There are clear benefits for concentrating on harmspots for location-based experiments rather than hotspots, as shown in the following sections.

As a measure of harm, we can rely on a metric developed by Sherman (2007, 2013; Sherman, Neyroud, et al., 2016), who suggested assigning weights to each classification of crime according to the sentencing guidelines available. This conversion of the number of crime counts into crime harm requires multiplying the number of crimes by the minimum number of days in prison that the crime would attract if an offender were to be convicted. A focus on the concentration of harm is more likely to result in greater good to society, as it will focus the police primarily on violence and crimes against persons. One homicide prevented is substantially more beneficial to society than preventing a few, or even many, bicycle thefts. By targeting harmful places, these 'more serious' events are more likely to be prevented because they are more likely to be targeted accurately than through a focus on crime count concentrations. Operatively, this suggests a focus on crimes against persons.

Notably, harmspots can be 'cooled down' when police conduct high-visibility patrolling, particularly on foot. Foot patrols have always been one of the most fundamental tasks of policing, being both effective and efficient (Ratcliffe et al., 2011). This is the cornerstone of deterrence and preventative measures, especially when conducted in crime hotspots during peak times. Rigorous research conducted over the past 25 years clearly demonstrates that directed patrols in hotspots reduce crime and antisocial behaviour (Braga et al., 2012). More importantly, patrols prevent crime and antisocial behaviour instead of reacting to them after they have already occurred.

Temporal-based randomisation

As a final alternative, we can use time as a means to group together units and assign them to treatment and non-treatment groups. One specific example is the use of police shifts (e.g. 08.00–17.00 shift) as the unit of randomisation and analysis (see Ariel et al., 2015; Ariel & Farrar, 2012). One clear benefit is sample size: even in a department with say 100 front-line officers, there are thousands of shifts every year. When designing an experiment on the effect of Tasers in policing, we can capitalise on this factor. We can randomly divide all shifts into treatment shifts and control shifts: during treatment shifts, all officers equipped with Tasers will conduct their regular patrols, and during control shifts, all officers will not be equipped with Tasers but will still conduct their regular patrols. At the end of the trial, we can compare and contrast the rates of use of force in the treatment versus control shifts.

Randomly assigning shifts as the unit of analysis is not *ideal*, given the potential contamination effect on other units, but it may be the most *optimal* unit for the policing context (and has applications to other shift-based work cultures, particularly emergency services). The issue with contamination when using shifts is as follows: the same officers experience both treatment and control shifts, so there is the likelihood that behavioural modifications due to treatment conditions can be 'carried over' into control conditions. If Tasers affect behaviour, then there may be a learning mechanism at play, where officers adapt their overall behaviour (and possibly attitudes), and this broader change affects them during control conditions as well. We might speculate that as officers begin to habituate their modified behaviour, they adopt the 'Taser version' of themselves whenever on patrol.

In such an experiment, officers participating on multiple occasions in both treatment and control conditions potentially violate the independence of the treatment and control groups (Rubin, 1990a) and the requirement that observations be independent (see Ariel et al., 2015). However, the unit of analysis in this randomisation procedure is the *shift*, not the officer. The set of conditions encountered during each shift cannot be repeated because time moves only in one direction. The manipulation will be whether the *shift* involves Tasers or no Tasers. The use-of-force outcome is then driven by officers with Tasers during each shift, versus shifts without Tasers. Likewise, because the shift was randomised, and officers experienced multiple shifts with and without Tasers, we know that, *on average*, all else was equal, including which officer was involved.

Similarly, being able to define units, treatments and outcomes in this detailed way means that we can be surgical about where violations to the experimental protocol are occurring. More importantly, however, spillover effects often result from experiments and, indeed, may *be* the intention (Angelucci & Di Maro, 2016). In our tests on body-worn cameras, for example (see below for more details), officers were exposed to both treatment and control conditions: the spillover meant that officers in control conditions were affected by their counterpart treatment conditions and altered their behaviour regardless of treatment condition.

Put another way, the exposure of officers to both treatment and control conditions is likely to affect the estimation of treatment effects, we think, asymmetrically. That is to say, officers in control shifts are likely to change their behaviour as a result of exposure to Tasers during treatment shifts. In the absence of detailed evidence, the working hypothesis is that during control shifts officers would change their behaviour to mimic that of treatment shifts. The spillover would therefore act to shrink the gap between treatment and control conditions by making control shifts more like treatment shifts. If true, this means that the estimated effect represents lower bound estimates of the intervention effect, rather than inflating the error. In other words, this so-called flaw makes our job in showing a significant difference from control conditions even harder, not easier, than a more conservative statistical test.

A final word regarding the choice of units of randomisation

The protocol for analysing experiments requires the experimenter to decide ahead of time the exact formulation of the units of analysis. The axiom in different experimental disciplines, particularly those with more elaborate designs compared to social science experiments, is that 'you analyse them as you have randomised them'. This general rule, set sometime ago by R.A. Fisher (cited in Boruch, 1997, p. 195), is fundamental: a trial, which is primarily concerned with the causal link between the intervention and the outcome(s), must be grounded in a pre-specified proposition about the units under investigation.[9] N.R. Parsons et al. (2018) reiterated this rule by saying that 'the experimental unit is usually the unit of statistical analysis', illustrating that a deviation from the general rule can be problematic.[10] Pashley et al. (2020) summarise this issue with the following remarks:

---

[9]See more in Kaiser (2012, p. 3867).

[10]On the other hand, subgroup analyses may be a different case. As Rothwell (2005) summarised, 'Subgroup analyses are important if there are potentially large differences between groups in the risk of a poor outcome with or without treatment, if there is potential heterogeneity of treatment effect in relation to pathophysiology, if there are practical questions about when to treat, or if there are doubts about benefit in specific groups, such as elderly people, which are leading to potentially inappropriate under treatment. Analyses must be predefined, carefully justified, and limited to a few clinically important questions, and post hoc observations should be treated with scepticism irrespective of their statistical significance. If important subgroup effects are anticipated, trials should either be powered to detect them reliably or pooled analyses of several trials should be undertaken' (p. 176).

It is a long-standing idea in statistics that the design of an experiment should inform its analysis. Fisher placed the physical act of randomisation at the center of his inferential theory, enshrining it as 'the reasoned basis' for inference (Fisher, 1935). Building on these insights, Kempthorne (1955) proposed a randomisation theory of inference from experiments, in which inference follows from the precise randomisation mechanism used in the design. This approach has gained popularity in the causal inference literature because it relies on very few assumptions (Imbens & Rubin, 2015; Splawa-Neyman et al., 1990). (p. 2)

Thus, it is usually *not* legitimate to analyse an experiment randomised according to one design as if it had been randomised according to another, unless certain specific and strict conditions are met (Kempthorne, 1955). If the researcher randomly allocated offenders as 'units' into treatment and control conditions, then the 'units' of analysis should be the offenders, not their victims, and conclusions based on analyses with victims' data are not purely derived from the randomisation procedure.[11] The victim data interact with the offender data, but a conclusion about victims is not *directly* derived from the randomisation of victims but from the interaction term between victims and offenders. It is not a direct measure that results from the exposure of offenders to treatments; it is mitigated by the offenders. Similarly, in a study that randomly assigned police *shifts* into Taser/no-Taser conditions, the analysis of *offenders' conviction* data as a result of using Tasers cannot be considered a main outcome of the RCT (or an 'outcome variation' of the main effect of the RCT), but rather analysis from an alteration of the overall test design. The factors associated with the decisions to charge, prosecute and convict a person for assaulting an officer, for example, are elaborate. The organisational, psychological and environmental processes linked to these decisions are formed based on the characteristics of the offence and the offender, not the police shift. This analytical transition might be common (Ariel, Mitchell, et al., 2020), but the price we pay for this plan is a conversion of the RCT into quasi-experimental design (see Chapter 4 for further reading).

For these reasons, it should be clear why the choice of the unit of analysis is critical: there is no turning back once the trial has commenced. However, the issues are

---

[11]In practice, many suggest that cluster randomised controlled trials do not follow this rule, with one unit of randomisation (e.g., the classroom) and a unit of analysis (e.g., students' scores). Similarly, multiple informant studies, for example parent-child-teacher reporting, or a ripple-effect experiment, may also seem to deviate from the 'analyse them as you randomise them' idiom.

However, we must be careful not to confuse between data analysis and hypothesis testing: with clustered randomisation, we are observing two worlds: one in which the intervention exists, and a world in which it does not. Any unit of analysis available to measure must be connected to the treatment/no-treatment conditions .

compounded when a different randomisation protocol is used midcourse for the assignment of units, for example, a new treatment provider is used, or when an additional eligibility criterion is entered after the test has already commenced. These scenarios are undesirable but not foreign. They do, however, lead to additional costs, as the new conditions require that a new randomisation sequence be used. In practical terms, the trial restarts as well, as we begin 'counting' the outcomes from the moment a new randomisation protocol sequence is used. Thus, when the choice of units, or their characteristics, is altered from its original definition in the experimental protocol (see Chapter 5), it has a dramatic effect on the study. Particular attention is therefore required to selecting the most appropriate or pragmatic units ahead of time.

To add to the complexity, there are no generalised practical, theoretical or statistical guidelines to help make the decision about which unit would be ideal given the study's research questions. Common practice is to replicate previous experimental protocols or make pragmatic choices depending on study parameters. For example, optimal experiments on hotspots policing should use street segments (Weisburd et al., 2012), unless there are good reasons to deviate from this rule of thumb. Still, this is not a hard rule and may vary from study to study (see Braga, Weisburd, et al., 2019). The point to emphasise is this: choose your units wisely.

Should we use significance tests for baseline imbalances after randomisation?

> Performing a significance test to compare baseline variables is to assess the probability of something having occurred by chance when we know that it did occur by chance. Such a procedure is clearly absurd. (Altman, 1985, p. 126)

The short answer to the question posed in this title is 'no'. The primary reasons why we do not have to conduct statistical tests for the pre-experimental balance are as follows.

First, the statistical tests that are used to analyse the results of the experiment already consider the possibility of non-equivalent conditions occurring by chance (Senn, 1994; see also Mutz et al., 2019). The post-test statistical significance testing allows for the possibility that pre-test scores of some covariates may fluctuate enough to produce significant differences between the experimental arms.

Second, with a sufficient number of statistical tests, one or more statistical comparisons is highly likely to emerge as statistically significant, by chance alone. With $p < 0.05$, it is expected that at least one comparison would yield a statistically significant difference – or one in 20 tests. Therefore, we know that tests may return results of significant differences, no matter what the true population covariance means (see also Bolzern et al., 2019, in the context of cluster RCTs).

Third, experimenters are cognisant that these baseline differences are inevitable, especially when the randomised groups are small (Chu et al., 2012). In the long run, however, these differences will average out. With repeated tests on the same phenomenon, the baseline means of the covariates in the two groups will overlap. Random allocation will create, over multiple tests in which participants are randomly allocated into treatment and control conditions, conditions similar to pretest levels. This feature of synthesised random assignments from the same population over repeated occurrences is extended to any particular variable, or group of variables, and the ways in which they interact. Thus, if we trust random allocation, which is underpinned by probability theory, then whatever baseline differences may occur due to chance will cancel out over time, so they can largely be ignored.[12]

As we discuss in the box below, the consequence of these assumptions is that there is no need to prove that some factors are significantly more pronounced in one group over the other, because we already assume the probability that these differences will emerge. As a result, we do not need to control for any covariates, which some statisticians have convincingly argued based on several grounds (Altman & Doré, 1990; De Boer et al., 2015). First, significance testing and statistical modelling as a guide to whether baseline imbalances exist can be viewed as inappropriate (Senn, 1989, 1994, 1995). As Altman (1985) pointed out, these tests can show whether the observed differences happen by chance, though we already know that any observed significant differences could be due to chance.

Furthermore, baseline equality needs to be a derivative of logical procedures, not of a parametric formula (Peduzzi et al., 2002). This means that baseline characteristics should be compared using clinical and 'common sense knowledge' about which variables are important to the primary outcome, and only then adjusting for their known effect through design, if necessary. If there are important covariates, they should be included in the process of random allocation before the experiment commences – a topic we cover below. Assessing baseline imbalance should be descriptive and transparent, but there is no need to 'add' equality between two groups postrandomisation – a view that was in fact embraced in the Cochrane Collaboration Guidelines and Consolidated Standards of Reporting Trials, which we

---

[12]We urge the reader to read the superb review and critique of Fisher's view that randomisation 'relieve[s] the experimenter from the anxiety of considering and estimating the magnitude of the innumerable causes by which the data may be disturbed' by Saint-Mont (2015).

discuss in Chapter 5. These tests are not only unnecessary but also can be counter-beneficial. Balance testing is inappropriate and, as remarked by Mutz and Pemantle (2012), conducting balance tests

> demonstrates a fundamental misunderstanding of what random assignment does and does not accomplish. A well-executed random assignment to experimental conditions does not promise to make experimental groups equal on all possible dimensions or on any one characteristic, or even a specified subset of them. . . . Doesn't this lack of across-the-board equivalence pose problems for drawing strong causal inferences? Contrary to popular belief, it does not. This idea was the single fundamental scientific contribution of R.A. Fisher. It is not necessary for experimental conditions to be identical in all possible respects. (p. 5)

Inferential statistics and probability theory

We emphasise that statistical significance is largely driven by sample size, so the larger the sample size, the more likely we are to find statistically significant differences. For example, we might observe a difference of 3.9% between treatment and control groups on a particular pre-intervention measure. With a sample of 200 participants, that difference may not be 'significant' at the 5% level, but with 2000 participants, it is much more likely to be 'found' to be significant. Using an independent samples t-test to calculate the differences between two means (M) illustrates this point. If the first group's $M = 5.10$ (standard deviation [SD] = 1.20) and the second group's $M = 5.30$ ($SD = 1.10$), then the difference between the two means is statistically significant when each sample contains 1000 participants ($t(1998) = 3.88$, $p < 0.0001$), but it does not reach accepted statistical significance with 100 participants in each group ($t(198) = 1.229$, $p = 0.22$). In these two scenarios, the percent difference between the two groups is the same – 3.9% – but only the test with 2000 participants overall will yield a significant difference.

Similarly, the more statistical tests we conduct, the more likely we are to find 'significant' differences by chance. With 20 variables (or covariates, as the case may be), we would expect to find that one of these variables is different (e.g. more or less of 'it' in one experimental group over the other) by chance, using a benchmark of 5% significance (1-in-20 = 5%). However, the problem with *statistically testing* so many variables is much worse. The 5% significance level is our benchmark for *Type I* errors – the probability of falsely rejecting the null hypotheses and concluding that there is a difference between the two groups when in fact there is no such difference (i.e. nil effect) – in terms of the variable in question. When we are independently testing many variables for differences, such as is suggested when looking at 'baseline equivalence', we might conclude that there is a difference with a much greater probability than the 5% we think we are using (with 20 variables, it would in fact be a 64% chance of falsely rejecting the null hypothesis).

This is often described as the problem of running *multiple outcomes testing* (see discussion in Langley et al., 2020), and an illustration helps to understand the issue. With M independent tests and an acceptable error rate of 5% (α), the formula for calculating the probability of falsely rejecting the null (no difference) hypothesis is $1 - (1 - \alpha)^\wedge M$. So with $M = 1$, the result is $1 - (1 - 0.05)1 = 0.05$ (5%) as expected, with a 5% significance (α) level. Nevertheless, with $M = 2$, the chance of falsely rejecting the null in one of the tests increases to 10%, with $M = 3$, it is 14%, with $M = 4$, it is 19% and with $M = 5$, it is 23%. This means that with five covariates tested, there is a 23% chance of rejecting the null hypothesis and incorrectly concluding that there is a difference where there is none.

Instead of significance testing, we advocate comparing the means and distributions of baseline variables graphically and appraising any differences in terms of whether they may have a meaningful influence on outcomes. Imagine a situation where we could only use simple randomisation and not assess balance until the end of the trial. We then find that the two groups differed in their means by 3% at the pre-intervention stage. When we look at their post-intervention outcome, we see a difference of 10%. In that situation, we might conclude that the 'gain' of 7% is because of the intervention and that the pre-intervention difference is not sufficient to explain the outcome observed. (In reality, we would include the pre-intervention data in our analysis, so we have the 'baseline controlled' model.)

Box 2.3

Common Tests for Baseline Equivalence in Non-RCTs

What about quasi-experimental designs (see more in Chapter 4)? It is common to use statistical tests to assess the degree to which the analysis created baseline balance and to fix it in case there is no balance. There are many statistical tests available for assessing baseline imbalances in quasi-experimental research, including the use of t-tests, chi-squares and other tests for comparing their baseline inequality (Stuart & Rubin, 2008). The applications of many of these tests are covered in this series. The choice of test depends on the level of measurement of the variables (i.e. continuous, binary, time-dependent etc.; see review in Trowman et al., 2007; Wei & Zhang, 2001).

Randomisation and sample size considerations

We have noted repeatedly that in order for randomisation to create pre-experimental balance, we need to have a sufficiently large study. How many participants would we need? This is a difficult question to address, and often experimenters are concerned about having 'just enough' cases. As sample sizes increase, the absolute size of imbalance should be reduced, owing to a reduction in sampling error (Roberts & Torgerson, 1999), while the likelihood for an equal baseline increases as sample size increases (Lachin, 1988b). As the sample size increases, the effect of baseline covariates should also become negligible, including the effects of unbalanced data variance (Berger & Weinstein, 2004, p. 516). Still, what is the optimal number of units that would take into account these probability considerations, as well as the costs, availability and feasibility of the greater number of participants?

We know that the more units you have, the more expensive a study becomes. Managing a large study can be difficult, and maintaining consistency of treatment delivery, control over the allocation of the treatments and data collection are all equally costly (see Gelman et al., 2020; Nelson et al., 2015; Weisburd et al., 2003). Therefore, real-world scientists are always happy to reduce the cost of the experiment by identifying the *minimum* number of participants they would have to recruit for a trial (see such considerations in Hayward et al., 1992).

When the properties of random assignments are met, the randomisation is expected to create two groups that have (1) the same size and (2) the same 'baseline covariance' (Beller et al., 2002; Lachin, 1988a; Pocock, 1979; Torgerson & Torgerson, 2003).[13] We explain these two qualities below in the context of sample size.

(1) *Group size balance:* Consider the intuitive approach of having 100 participants in each of the experimental groups. Lachin (1988a) showed that a study with exactly 100 participants in each group has about a 5% chance to have a 60:40 split in terms of the number of units in each group. Lachin (1988b) furthered this concern by plotting the chance of treatment imbalance as a function of sample size for different fractions of the total sample size for larger treatment groups. The exercise shows that for harmonic *n* (i.e. in both groups) of 386, 96, 44 and 24 participants, there is a 5% chance of imbalance of 55%, 60%, 65% and 70%, respectively (see also Chow & Liu, 2004, p. 133). For example, statisticians have shown that the assumption of equality (i.e. obtaining a 50:50 split) is 'very likely' when the study comprises 1000 participants or more (Beller et al., 2002; Yusuf et al., 1984). In other words, while there is always the possibility of a freak anomaly that even a study with 1000 participants will

---

[13]The question of sample size is linked to 'statistical power', an issue with which social scientists like David Weisburd are particularly concerned (Weisburd & Gill, 2014). We return to this more formidably in Chapter 4.

end up in baseline inequality in terms of group sizes, it is very unusual and highly unlikely to get anything other than a result close to 50:50 split when the sample size is this large.

(2) *Pre-randomisation equilibrium:* Covariates are the important characteristics that define the participants – for example, age, gender, ethnicity or criminal background. When covariates are found to be different between the experimental arms before the administration of the intervention, it means that the distribution of pre-treatment variables can influence the outcome (Berger, 2004), because baseline imbalances create groups that are not comparable. Any apparent effect of the intervention may be spurious and lead to biased parameter estimations (Altman, 1996). Consider a hypothetical trial with two groups, where some covariance variable exists in 15% of the overall sample (e.g. Kernan et al., 1999, p. 20). The chance that the two groups will differ by more than 10% for the proportion of participants with the covariance variable is 33% for a trial of 30 patients, 24% for a trial of 50 patients, 10% for a trial of 100 patients, 3% for a trial of 200 patients and 0.3% for a trial of 400 patients. Thus, it is commonly understood that as the sample size increases, the effect of baseline covariates should become negligible. The larger the study, the lower the likelihood that the covariates will be distributed unequally due to chance (i.e. random assignment).

As another example, assume that a variable of interest occurs in 20% of the sample (e.g. 20% of the patients will have an allergic reaction to the treatment, or 20% of the offenders will drop out of the study once they are offered the treatment). In a sample of 200, this variable will exist in 40 individuals. If the split between treatment and control due to random assignment was a perfect 50:50, then you would expect 20 participants in the treatment group and 20 participants in the control group to experience that variable. Nevertheless, when the split is expected to be 60:40 (even for 'just 5%' of the time that you run this experiment), you may end up with 24 participants in one group versus 16 participants in the other group who have experienced this variable. The difference then grows to 50%.

The problem is that even if the assumption that larger samples lead to baseline balance were true, many field experiments are not large enough. Studies of more than a few hundred participants randomly assigned into treatment and control conditions are not common in criminology. Field experiments in the criminal justice system are usually much smaller than this, with relatively weak effects. In Farrington and Welsh's (2005) systematic review of 'high-quality' RCTs in criminology, for example, nearly one-third (29%) of the reported trials had sample sizes below 200. In Peter Neyroud's (2017) review of experiments in policing, just 12 of the 122 policing experiments identified reached a sample size of 1000. A longitudinal view of the growth of some experiments in criminology can be found in Braga et al. (2014), and while the trajectory of experimental criminology is up, the number of 'large trials' remains an exception.

Summarising the benefits of randomisation

Randomisation is beneficial for several reasons. Perhaps, most importantly, it breaks the link between treatment allocation and potential outcomes. With randomisation, we do not need to be concerned that differences between groups can be attributed to the manipulation (non-randomisation) of treatment allocation. In doing so, randomisation allows us to generate a comparison group that is 'as similar as possible' to our intervention group.

We have then tried to convey that as sample size is increased, randomisation is likely to create equal groups. This concern is of particular interest when the studied effect is suspected to be relatively small as well, therefore being more sensitive to 'disappear' in the noise that is created by the covariates and the outliers. Probability theory states that imbalances between the groups will become negligible as sample size increases, which refers to a distribution of replicated trials. We have then shown that there is still value in these studies, with baseline inequality that was not 'sorted' by the random assignment. Results from experiments compiled across a series of replications (e.g. those seen in numerous trials of police wearing body-worn cameras) would then 'cancel out' any differences that may have been detected between the groups in any one test of the treatment effect.

What does all of this mean? The good news is that if we are content that our randomisation was successful, then we are able to make our causal inference. That is, if we have randomised, then we know, by virtue of randomisation, that observed differences between groups arose because of the randomised group membership, rather than something else. If we have randomised properly, and in enough numbers, we can compare the outcomes for treatment and control groups and be sure that the difference between them is our causal estimate. What remains to be seen is how other features of the experiment fit in with the process of randomisation – a topic we turn to in the next chapter.

Chapter Summary

- In this chapter, we go deeper into the theory behind the 'magic' of randomisation – probability theory. We also discuss some of the different forms that random assignment can take, including methods for reducing pre-experimental differences using various techniques.
- First, we review the importance of comparisons and relative effectiveness in causal research. We then discuss common procedures of randomisation that help researchers achieve baseline balance between treatment and comparison groups.

- Finally, we discuss units of randomisation that experimenters can use to analyse cause-and-effect relationships, which is then followed by some more technical considerations such as sample size and whether we should test for baseline imbalances after randomisation.

Further Reading

Morgan, S. L., & Winship, C. (2015). *Counterfactuals and causal inference.* Cambridge University Press.

Various authors have extensively studied the issue of counterfactuals over the years, applying different approaches to conceptualising and measuring them. Researchers have developed a variety of methods for applying the counterfactual approach to observational data analyses, which has been only briefly touched upon in this chapter. This book offers further reading on these concepts, as well as information on the statistical instruments used for causal inference under observational conditions. Morgan and Winship's book provide one of the most comprehensive accounts of these approaches, published as part of a Cambridge University series on analytical methods for social research.

Altman, D. G. (1985). Comparability of randomised groups. *Journal of the Royal Statistical Society: Series D (The Statistician), 34*(1), 125–136.

One issue discussed in this chapter is the lack of awareness amongst scholars regarding the inappropriateness of applying inferential statistics to test for baseline imbalance in randomised controlled trials. This article provides further reading on this issue, from both conceptual as well as statistical perspectives.

3

C IS FOR CONTROL

Chapter Overview

What do we mean by control?

The second fundamental rule of causal research is to maintain a high level of control in the conduct of the study. This includes all the steps associated with experimental methods, from the early design stages up to the dissemination of the results. In a laboratory study, the environment is highly controlled, as per the quote from Cox and Reid (2000) that opens this book. Of course, the strength of the control is always relative to the context in which the experiment is conducted, and the purity of the experimental design is often defied by 'human error', such as post-randomisation decisions that change the course of the experimental blueprint. Participants drop out, treatment providers often challenge the treatment fidelity by delivering the intervention inconsistently (e.g. police–public interactions do not all look the same) and random allocations are frequently broken. Real-life constraints are the price we pay for experiments conducted in the social sciences.

Subsequently, these issues are critical because an error can jeopardise the integrity and the overall validity of the test. Careful planning is necessary by taking into account the potential violations of certain rules. There are always gaps between the experimental protocol and the experiments researchers administer in practice. Still, we must be mindful not to overly compromise, otherwise certain experiments will be 'doomed to failure' due to lack of controls before they even commence.

Experimenters need to have a detailed plan on how to increase their control over the experiment. In this chapter, we discuss two core issues[1]:

1   Threats to the internal validity of the test
2   Threats to the external validity of the test

Threats to internal validity

One of the most important reasons to keep controls at reasonable levels in the experiment is to reduce threats to the **internal validity** of the test. These threats have been known to scholars for some time (D.T. Campbell & Stanley, 1963); however, they are often overlooked. Criminology is no different, and while the fundamentals of science are violated regularly in many fields, some evaluation studies are implemented despite being based on poor designs that jeopardise internal validity.

Internal validity was well described by the popular social research methods website Research Methods Knowledge Base (Trochim, 2006):

---

[1]For a detailed and more comprehensive review, see Shadish et al. (2002). See also Bryman (2016).

Internal validity is only relevant in studies that try to establish a causal relationship. It's not relevant in most observational or descriptive studies, for instance. But for studies that assess the effects of social programs or interventions, internal validity is perhaps the primary consideration. In those contexts, you would like to be able to conclude that your program or treatment made a difference - it improved test scores or reduced symptomology. But there may be lots of reasons, other than your program, why test scores may improve or symptoms may reduce. The key question in internal validity is whether observed changes can be attributed to your program or intervention (i.e. the cause) and not to other possible causes (sometimes described as 'alternative explanations' for the outcome).

We prefer this definition of internal validity due to its practical focus – the factors that jeopardise our conclusion that the intervention is the cause of the differences between the group that was exposed to the treatment and those that served as controls. Making a claim that a treatment affected the participants in some way is a powerful statement, and we must place sufficient controls over the execution of the experiment so that this finding cannot be attributed to some other factor(s).

The principal origin of discussions of validity, then, is the American Psychological Association's *Technical Recommendations for Psychological Tests and Diagnostic Techniques* (1954), although the term *validity* was already used in and exemplified by Thurstone's handbook *The Reliability and Validity of Tests: Derivation and Interpretation of Fundamental Formulae Concerned With Reliability and Validity of Tests and Illustrative Problems* (1931) and R.A. Fisher's 1935 *Design of Experiments*, which explored in more detail the design of the experiment, including validity. As carefully reviewed by Heukelom (2009), the *Technical Recommendations* manual did *not* include the concepts internal and external validity.[2] It was only when Donald Campbell (1957) challenged the focus and emphasised the distinction between internal and external validity, and then fully developed it with Stanley in their premier reader *Experimental and Quasi-Experimental Designs for Research on Teaching* (D.T. Campbell & Stanley, 1963):

Validity will be evaluated in terms of two major criteria. First, and as a basic minimum, is what can be called internal validity: did in fact the experimental stimulus make some significant difference in this specific instance? The second criterion is that of external validity, representativeness, or generalizability: to what populations, settings, and variables can this effect be generalized. (D.T. Campbell, 1957, pp. 296-298)

Thus, internal validity is the 'basic minimum without which any experiment is uninterpretable: did in fact the experimental treatments make a difference in this

---

[2]The American Psychological Association's *Technical Recommendations* manual (1954) distinguished four types of validity: (1) content validity, (2) predictive validity, (3) concurrent validity and (4) construct validity.

specific experimental instance?' (D.T. Campbell & Stanley, 1963, p. 5). In later years, D.T. Campbell (1968) referred to this type of validity as 'local molar causal validity', which emphasises that 'causal' inference is limited to the context of the particular 'treatments, outcomes, times, settings and persons studied' (Shadish et al., 2002, p. 54). The word *molar*, which designates here the body of matter as a whole, suggests that 'experiments test treatments that are a complex package consisting of many components, all of which are tested as a whole within the treatment condition' (Shadish et al., 2002, p. 54). In other words, local molar causal validity – internal validity, in short – is about whether a complex and multivariate treatment package is causally linked to a change in the dependent variable(s) – in particular settings, at specific times and for certain participants (see also Coldwell & Herbst, 2004, p. 40).

We should emphasise that the treatment is always complex, often like a 'black box', because most interventions in the social sciences are complex and multifaceted. For example, when cognitive behavioural therapy (CBT) is found to be an effective intervention for offenders (e.g. see Barnes et al., 2017), it naturally includes multiple elements that makes the CBT stimulus 'work', and it is difficult to pinpoint a singular element as the most important component. For this reason, it is vital for experimenters to report precisely and with as many details as possible what the intervention included, not only to assess its content validity – and its descriptive validity (C.E. Gill, 2011) – but also to understand what it is about 'it' that caused an effect.

Box 3.1

Inter-Group Balance of Baseline Variables

Inter-group balance in terms of baseline variables is primarily about the internal validity of the test and the foundation of relative change in causal research. Let us take an example as a case in point: does the integration of closed-circuit television (CCTV) with proactive police activity generate a crime-control benefit compared to no-treatment conditions (Piza et al., 2015)? What is the effect of CBT on the recidivism of adult and juvenile offenders (Landenberger & Lipsey, 2005)? Do court-mandated batterer intervention programmes for domestic violence reduce violence (Feder & Wilson, 2005)?

To answer these questions, scholars need to compare these interventions to something else (the aforementioned counterfactuals). However, as we explained in Chapter 2, if the groups are not comparable at baseline, then a causal relationship can become difficult to determine. In order to show that using active CCTVs has a positive effect on police and crime outcomes, locations with CCTV must be compared to locations without CCTV – however, the locations must be similar. To illustrate how CBT reduces recidivism requires a group of offenders who have undergone this type of treatment and a comparison group of offenders who have not, the two groups of offenders need to be balanced – that is, the same. If the groups are not the same *before* the treatment group was exposed to the intervention (and, again, the comparison group did not have the intervention), then there can be serious challenges to the causal estimate – the internal validity of the test. If the CCTV systems are installed in low-crime areas and the control locations in high-crime areas, then the differences the researchers have found could easily be attributed to the crime levels, rather than the CCTV systems. Similarly, studies that have shown how court-mandated domestic abuser intervention reduces the reporting of repeated abuse to the police must have similar offenders in both groups – the group that received the intervention and the group that did not. Otherwise, we cannot be sure that these state-funded interventions reduce domestic violence.

The primary risk to internal validity is that we will conclude that variations in the dependent variables are attributable to the intervention, when in fact some other factor may be responsible. For example, if we have pre-test and post-test measures – that is, one observation of the baseline data before the intervention (e.g. last year's crime figures), and one more observation of the outcome data after the intervention (post-test crime figures) – changes from the pre to the post periods might be attributable to another factor. Under such conditions, we will not be able to conclude that the intervention caused the changes.

However, this risk is just one of several. The overall issue is that there are *systematic* rather than random variations in the differences between the treatment and the control conditions, and these differences between the two groups are the cause of the disparity, not the treatment itself. In attempting to break down this issue in detail, we turn to the factors most famously discussed and developed by D.T. Campbell (1957; see also Campbell & Stanley, 1963). These are events that threaten the internal validity of the test – and primarily challenge the conclusion that the variation from the levels of the dependent variable before the intervention was administered and its levels afterwards was due to the intervention, and nothing else. Without accounting for these factors – in effect ruling them out as plausible limitations – the effect of the experimental stimulus will be said to be confounded and the inferences made from the experiment misleading. Some of these are naturally controlled for due to the random allocation of cases into treatment and control conditions. However, others remain a threat to the internal validity, even though the study follows an RCT design protocol. In the following pages, we review these threats and some solutions to remedy these hazards.

Box 3.2

Threats to Statistical Conclusion Validity

Shadish et al. (2002) offer one of the most celebrated tables about statistical conclusion validity. These threats were given different titles and references over the years, and the list can be reduced or increased. However, the list highlights the plethora of risk when trying to conclude that two variables are associated. Note that these threats supplement the issues we discuss here about internal validity.

Threats to statistical conclusion validity, or reasons why conclusions based on statistical analyses may be incorrect:

1 *Low statistical power:* An insufficiently powered experiment may incorrectly conclude that the relationship between treatment and outcome is not significant.
2 *Violated assumptions of statistical tests:* This can lead to overestimating or underestimating the size and significance of an effect.
3 *Fishing and the error rate problem:* Repeated tests for significant relationships, if uncorrected for the number of tests, can artificially inflate statistical significance.
4 *Unreliability of measures:* This weakens the relationship between two variables and strengthens or weakens the relationships among three or more variables.
5 *Restriction of range:* Reduced range on a variable usually weakens the relationship between it and another variable.
6 *Unreliability of treatment implementation:* If a treatment that is intended to be implemented in a standardised manner is implemented only partially for some respondents, effects may be underestimated compared with full implementation.
7 *Extraneous variance in the experimental setting:* Some features of an experimental setting may inflate error making detection of an effect more difficult.
8 *Heterogeneity of units:* Increased variability on the outcome variable within conditions increases error variance, making detection of a relationship more difficult. The more the units in a study are heterogeneous within conditions on an outcome variable, the greater will be the standard deviations on that variable and on any others correlated with it. Possible solution is blocking – measure relevant respondent characteristics and use them for blocking or as covariates.
9 *Inaccurate effect size estimation:* Some statistics systematically overestimate or underestimate the size of an effect.

History

These are the 'life' events occurring between the first and the second measurements (i.e. before and after participants are exposed to a stimulus). When history effects are controlled for, then we would conclude that extraneous effects are not responsible for the differences in the before and after measures. Researchers therefore aim to use measures that are unaffected by real-life conditions that may cause variations between the pre and the post measures, as opposed to the intended effect of the treatment under investigation. In field settings, however, external life events do occur and could therefore cause the participants to behave differently between the pre and the post observations, independent of the tested intervention.

It is often argued that when the treatment and the control groups of participants are *equally* exposed to historical events, then the threat to internal validity – that is, the causal relation between the treatment and the outcome – is mitigated. If all participants are affected equally by external factors – such as the weather, events reported in the news or a new chief executive officer – then all units are exposed to the same experiences (i.e. a fair playing field for both treatment and control groups). Therefore, as all units in the experiment go through the same historical processes and stimuli, then from an internal validity perspective we are less concerned.

It would be dangerous to assume, however, that the *balance* in the history effects that the two study groups experience is naturally maintained. First, the treatment and control study participants may be exposed to these external events differentially. This can be common in policing experiments, for example, in which batches of participants enter their assigned group sequentially (e.g. see Alderman, 2020; Ariel & Langley, 2019). Therefore, since history effects have a temporal component (they happened at *some* time but then stopped), the events may have occurred during the recruitment and assignment of one batch, but not in others.

Second, as we explained in detail in Chapter 2, randomisation does not always work. By chance alone, an experimenter may be confronted with 'more' or 'less' exposure to historical events in one group. Furthermore, the differences between the treatment and the control groups due to one group experiencing more of some type(s) of exposures than the other are more pronounced in smaller studies, when the effect of small yet influential outliers exposed to historical events is stronger. Under these settings, internal validity is at risk, despite the assumption of balanced exposure due to random assignment. The concern is exacerbated with types of historical events that are unmeasured or unmeasurable (and therefore cannot be controlled), that due to chance alone are more pronounced in one group rather than the other.

The response to this realistic view of experiments expounds upon our earlier argument that the evolution of scientific knowledge takes place through an iterative process: brick by brick. No one would advocate a strong case for evidence-informed policy based on only one experiment. The more *replications* on a particular research question which point to the same direction and with a normally distributed set of effect sizes, the more we trust the results of the original test. In a similar way, if there are historical events that jeopardise the internal validity of *one* experiment, we expect these extraneous effects to dissipate when considering the effects in a *series* of experiments.

Sherman and Weisburd's (1995) influential RCT on hotspots policing is a great example in this sense. It is conceivable that a handful out of the 115 crime hotspots that were randomly assigned into *one* of the experimental arms (treatment or control conditions) experienced gentrification, a new local neighbourhood watch scheme or other historic effects that may have caused variations between the before and after periods, independent of the treatment. The differential experience of these hotspots withstood the desired balance due to random allocation. However, the more experiments we have that replicate Sherman and Weisburd's (1995) study, the more we are convinced by their fundamental hypothesis and their findings (see Braga, Turchan, et al., 2019): placing 'cops on the dots' reduces crime compared to hotspots where such proactive and visible policing is not present. The more data points on a forest plot – that is, a graph that synthesises the results from several experiments – the more we are assured that the result is consistent, and the less we are concerned by potential historic events that may have affected the results in one group more than the other, between the pre and the post measures of the experiment.

Therefore, while we *should* assume that historic events are spread evenly between the treatment and the control groups, we *suspect* that they might occur, especially in smaller or more complex trials. Given this concern, we should always take great care in sound planning and proper protocol implementation, with tight controls over the entire experimental process.

Maturation

This term refers to internal processes that the participants go through over the passage of time, between the first and the second measurements, and are responsible for the before–after variations, notwithstanding the studied treatment effect. These may include growing older, gaining new knowledge, making new peers, experiencing 'turning point' events and so on. The changes occur 'naturally' without the treatment effect, and when they occur, we may find it difficult to conclude that the post-treatment observations are due to the intervention alone.

For example, if we take a group of offenders who are in their 30s, then any intervention to stop their criminal behaviour will have to perform 'above and beyond' the natural 'dying out' of crime phenomenon that the overwhelming majority of people experience as they reach their late 30s. The age–crime curve, which has been heavily studied in criminology, repeatedly shows that as people get older, they are substantially less likely to exhibit recidivism (Farrington, 1986; Steffensmeier et al., 1989; Steffensmeier et al., 2020). There are different theories for this phenomenon; however, the evidence is robust: the overwhelming majority of criminals do in fact step away from a life of criminal behaviour, or at the very least substantially reduce their involvement in crime (Hirschi & Gottfredson, 1983; Piquero & Brezina, 2001; Sampson & Laub, 2003, 2017; Sweeten et al., 2013). This is great news for society; however, it challenges our ability to evaluate an intervention designed to assist mature offenders to stop committing crimes. Over time, the participants would exhibit a reduction in criminal involvement versus the pretreatment measurement scores *anyway*, and we would find it difficult to ascribe the change to the treatment, rather than the natural fade out of crime patterns.

To be sure, maturation is not just an issue for longitudinal studies, or experiments that have an extensive follow-up period. Short-term experiments can suffer the same maturation issue – for example, in terms of reduced aptitude or ability over time; some participants naturally get more tired, agitated, annoyed or just generally under-perform between the first and the second measures (see review in Guo et al., 2016; Porter et al., 2004; A. Sutherland et al., 2017).[3] For instance, we may provide police cadets with 'procedural justice' training to ensure that they are more trusted by members of the community (greater legitimacy). If we conducted a before–after measurement of their knowledge of procedural justice – once before the training and then again after the training – then we might be measuring changes that the cadets have gone through as part of their *overall* personal development as cops. This maturation – natural change that would occur even in the absence of the procedural justice training – can threaten the internal validity of our test (Antrobus et al., 2019).

One common way to reduce maturation threat is by recruiting participants simultaneously, of generally the same age, gender and geospatial parameters. Remember, the allocation of units into treatment and control groups using randomisation is meant to dramatically reduce any differential due to chance. Random assignment is hypothesised to create balance between the groups of maturation as well. However, due to chance, this may not happen, especially in smaller studies. For this reason – just like we have shown for the effects of historical events – the more homogeneous the participants are at pre-allocation stage (in terms of age, gender etc.), the less likely that the treatment and control groups would 'mature' differently from one another.

Testing

Testing threats are the effects of being exposed to the test in the first measurement on the scores of the second measurement. The most intuitive example is

---

[3]The issue of appropriate follow-up periods is discussed often in the literature (e.g. D.T. Campbell & Stanley, 1963, p. 31; Farrington & Welsh, 2005), but there is not yet an agreement on what constitutes a valid period to follow up on cases post-allocation.

taking an exam and then taking the same exam again: the repeated exposure to the test, rather than any manipulation that came into play between the two tests, causes variations in the subsequent scores. For example, McDaniel et al. (2007) have shown that quizzing without additional reading improves performance on the criterial tests relative to material not targeted by quizzes. People are also affected by the test score itself and can be motivated to change behaviour based on these scores – so that the second time they are tested, they exhibit a change due to the first scores. For example, overweight individuals can be affected by their weight measurement and as a result embark on a new diet regime – with or without any intervention in place. This suggests that if an intervention is being tested at the same time as the participant's own initiative to lose weight, we would not be able to separate its efficacy from the participant's own volition (Shuger et al., 2011).

This issue is particularly pertinent with control participants. If the point of the pre-test measure is to create a baseline for both treatment and control participants, and then the measure itself becomes an active manipulation, we are biasing the study in favour of the control group – thus eliminating the no-treatment status of the participants. Suppose a study hypothesises that when police detectives have a 'warning chat' with prolific and serious offenders, the offenders would reduce their criminal behaviour (Ariel, Englefield, et al., 2019; Denley & Ariel, 2019; Englefield & Ariel, 2017; Frydensberg et al., 2019). Offenders are believed to be deterred by these warning conversations because they assume that the risk of apprehension has just been elevated (and no rational actor wishes to be caught). Now, suppose that the measure of interest is self-control, given the vast literature on the relationship between low self-control and crime (Gottfredson & Hirschi, 1990). If we measured all of the participants' self-control levels before the study, then exposure to the test itself may affect the participants' criminal behaviour. *However*, this is a more concerning issue with the control offenders who have not been contacted by the police, because the exposure to the pretest measure can potentially make them think that in fact the police have placed them under greater scrutiny as well – thus leading them to exercise self-control. Therefore, instead of having a clean comparison between treatment offenders who have been exposed to the warning chats and control offenders who had not been exposed to these warning chats, we are in fact comparing two 'treated' groups. While it is still the case that the participants exposed to the warning chat are in a 'better' position to reduce their criminal behaviour, the comparison is not clean. One could also argue that even the treatment group has been exposed to a pretest 'intervention' by taking a pre-intervention test (e.g. raising their awareness of their inappropriate lack of self-control), which therefore masks the true treatment effect.

However, the bias is more pronounced in the no-treatment group. The self-control test (Duckworth & Kern, 2011) would undermine the no-treatment status of the control group by serving as a stimulus – which by implication undermines the no-treatment status of the control group when they take the test again at the post-treatment stage.

Experimenters do not pay sufficient attention to these testing effects (Ariel, Sutherland, & Sherman, 2019; Willson & Putnam, 1982). There are some techniques that could be used to minimise the effect of testing – for example, by increasing the temporal lag between pre-test and post-test measures – however, these solutions are far from ideal. The most accurate and scientific way of ruling out the effect of testing is the Solomon four-group design (Solomon, 1949), which is discussed in Chapter 4; this is a design in which some participant groups are measured at pretest and others are not, in order to measure the potential effect of the pretest itself on the post-test. However, the Solomon four-group design is rare – and for good reason – as it is an elaborate design that requires a relatively large number of units and tighter control (Dukes et al., 1995; Pretorius & Pfeifer, 2010).

Instrumentation

Instrumentation threats derive from changes in the measurement instrument between the two measurement points. For example, researchers may use replacements of original observers or different coding techniques for calibration of the collection tool – so that the second measure becomes invariably different. This scenario suggests that even in the absence of the intervention, there are variations in the scores over time, due to using a different instrument, which makes it difficult to 'see' the treatment effect. Classic examples in the criminal justice are (a) variations in perceptions of crime severity over time (Apospori & Alpert, 1993), (b) amendments to the definition of a crime (e.g. moving a particular drug from class A to class B, or the definition of sexual assaults; Von Hofer, 2000), (c) deployment of more experienced students to conduct ride-alongs with the police to measure procedural justice practices (as the students gain practice in the observations procedure; Raudenbush & Sampson, 1999) and so on. In all of these examples, differences between the pre-test and the post-test measures are due to the variation in the instrumentation, rather than a result of the tested intervention – hence causing differences between the pre and the post phases, independent of the intervention.

It should become immediately clear why instrumentation is a greater threat in longitudinal experiments where multiple measurements take place, and often over a long period of time (C.E. Schwartz & Sprangers, 1999). As time passes, there is greater likelihood that changes will take place in terms of the instrumentation used in the study. Any study that evaluates an intervention over years is susceptible to this threat to the internal validity of the test, as necessary amendments to the instrument need to be made. For example, a study that is based on surveys of victims of crime must modernise the definition of certain victimisation types, not least due to legal changes but also in terms of what people are willing to share in these surveys. The case of fraud is a good example: while today it is clear that this crime type must be incorporated in victim surveys, since it is the most prevalent crime category in most metropolitan Western cities, for many years fraud was not included in victimisation surveys. The Crime Survey of England and Wales is an example of this: only recently were questions about fraud and particularly cyber-enabled fraud incorporated (Jansson, 2007). Subsequently, comparing the different surveys over the years – or between countries – becomes more challenging.

To clarify, instrumentation is different from *testing* effects because the latter refer to changes that occur in the participants themselves whereas instrumentation effects refer to changes in the data collection tool. *Testing* implies that the *participant* is getting better (or worse) through exposure to the instrument, and therefore their scores in the post-test are different from their scores in the pre-test as a result of the exposure to the pre-test measure. On the other hand, instrumentation implies that the change in pre and post scores is associated with the *method* of measuring the phenomenon. *Instrumentation* is therefore different from *maturation* as well, because *maturation* effects are linked to internal changes that the participant undergoes (independent of the experiment).

The practical implication of this threat to the internal validity of the test is to avoid (as much as possible) making changes to the instrument. D.M. Murray (1998) offers an elegant method of calibrating two or more instruments, if variations indeed have taken place – however, the need to adapt or to control for changes in the instrument is usually met with some resistance, because the results will always be somewhat suspect given this peril. One way to address this critique is to argue that when the treatment and the control group participants *both* go through the same variations in the instruments, then instrumentation effects are 'cancelled out' between the two groups. As we noted already, this is generally true. When the two groups both go through the same biases, then we can say that the two groups are equal (except for the exposure to the intervention by the treatment participants), which is one of the main priorities of experiments. Random assignment creates the same types and levels of concern that emerge with *instrumentation* effects. Still, we must assume that the effects are equally distributed between the two treatment and control groups – and this can be problematic in small experiments, as we argued earlier.

Box 3.3

Reactive Effects and Pre-test Measures

A related issue, raised by D.T. Campbell and Stanley (1963), deals with the reactive effect of testing, which we alluded to earlier when discussing internal validity. When participants are pretested – that is, measured before the administration of the intervention – they then may become more or less responsive to the treatment. There is also an external validity concern here in terms of the research settings, because the results obtained for a pretested population may be unrepresentative of the effects of the experimental variable for 'the un-pretested universe from which the experimental respondents were selected' (D.T. Campbell & Stanley, 1963, p. 6). This means that the pre-test observation affects the respondents, and therefore we are not looking at what the overall population might look like at baseline, rather than what the *observed* population might look like at baseline. The pre-test measurement has caused a change in the participants – which not only makes the intervention effect more pronounced or diluted – but also changes their natural settings and, consequently, the 'cleanliness' of unmeasured real-life settings.

In many experiments, the issue of the reactive effect of the pre-test measure is not a concern when the outcome measures (at baseline and post-treatment) are independent of the experiment – for example, based on records of delinquent behaviour in the facility in which prisoners are held (Gesch et al., 2002), or urine tests of drug addicts (Gordon et al., 2008). Arrests, summons, charges, crime harm and crime data are usually unaffected by the treatment, and therefore we should not expect a reactive effect. On the other hand, in some experiments, the reactive effect of the pre-test measure is often unavoidable because there is little that can be done if we want to measure variation over time. If we wish to show how participants' scores in any psychological battery or psychosocial test change between the pre-treatment and the post-treatment stages, than we must 'expose' the participants to a baseline test – as in any study that observes self-reported behaviour (e.g. Bilderbeck et al., 2013). Experimenters conducting these types of studies must acknowledge that, in fact, their baseline measure is not as 'clean' as it appears to be (see Linning et al., 2019) – which means that the external validity of the test is, by definition, a concern. The generalisation must be contextualised within the framework of observed baseline populations, rather than unaffected populations – unnatural settings.

Statistical regression to the mean

One interesting universal phenomenon of virtually all data trends is that over time they go up and down around a particular average. Think of the stock market, crime figures, public attitudes and so on – line graphs tend to fluctuate around a particular mean. In our context, this law of fluctuations around the mean suggests that the changes between the first and the second measurements may be attributed to these 'regressions' to the trend line, rather than the intervention. This is particularly concerning when we select a group of participants that is extremely higher or lower than such an average: the post-treatment measures may represent a natural return – that is, reversion – to the overall trend. In other words, patients, crime locations, offenders, victims or other units are selected from a population *because* they are significantly more affected by a particular problem. But they may regress to the overall mean *anyway*. If we see a drop in the post-treatment observation, then it could be that the group reverted to normal behaviour.

Several examples come to mind in the context of criminal justice research. Certain addresses may be chosen for police interventions, as they experience a disproportionate level of crime (Sherman et al., 1989); the top 'troubled families' are assigned to social care treatments because they are disproportionately more likely to have family members who are offenders, victims of crime or both (Hayden & Jenkins, 2014); hospitals apply 'triaging' techniques to solve overcrowding problems in emergency departments to fast track assault patients with less severe symptoms (Oredsson et al., 2011). The list goes on and suggests that many environments in which units are randomly assigned into treatment and control conditions are chosen based on extreme scores (low or high) and special needs. When these extreme scores are chosen – and by implication the participants who exhibit these pre-test scores – there is a risk that the scores would be less extreme in a retest of the original measure (see D.T. Campbell & Kenny, 1999). For instance, domestic violence offenders chosen for police intervention due to their potential high harm to their partners will not top the harm list in the subsequent year (Bland & Ariel, 2015, 2020), and businesses that experienced a great deal of vandalism will not be re-victimised with the same severity as prior to police intervention. Collectively, this phenomenon is ubiquitous.[4]

The regression to the mean phenomenon is particularly relevant for hotspots policing and experiments on the effectiveness of police presence in these small geospatial locations. We support the argument that the definition must incorporate a relatively extensive period of time for a place to be considered a hotspot – at least one year. If the location is not 'hot' on a year-to-year basis (or another long period of time), we should not call it a hotspot. Excluding places that experience a greater level of crime or harm, however, for a limited time – for example, a few weeks or a couple of months – does not mean that the police should not interject in order to 'cool' these places down. However, the intervention does not necessarily have

---

[4]For a historic review, see Stigler (1997), which shows that the concept was already discussed as early as 1869 by Francis Galton, who tried to understand why it was that 'talent or quality once it occurred tended to dissipate rather than grow' (p. 107).

to follow the prescribed treatment as suggested by Sherman and Weisburd's (1995) operationalisation of hotspots policing: a visible and deterring presence of officers in the hotspots for a period of approximately 15 minutes per visit (Barnes et al., 2020; Koper, 1995; Telep et al., 2014). The 'dynamic hotspots' approach, in which statisticians aim to predict the locations of the hotspots based on short-term models of near-repeat risk, has failed to find a significant effect on crime when assessed by independent researchers under controlled conditions (see review in Jaitman, 2018). These issues are important from an internal validity perspective because the police may target locations that are 'hot' with crime at the short-term level; however, these are locations that may 'cool down' anyway with the passage of time. Therefore, we will find it difficult to conclude that the police are responsible for reducing crime at these so-called hotspots; they simply will regress to the mean levels of crime the locations experienced at pretest. The selection of these areas based on high pre-test scores would be misleading, as these are not hotspots of crime.

To be sure, regression to the mean does not imply that the extreme scores exhibited in the pre-test would be reverted to the scores of the *overall* population mean. The crime hotspots will likely remain disproportionately higher in crime rates than 'coldspots' – however, it is also likely that, within the bounds of a 'hotspot', the particular locations would regress back to the standard crime levels that 'normal hotspots' experience. This is an important feature of regression to the mean: the means refer to averages that represent the population of units that are on the high-end tails of the distributions. The regression is not towards the overall population mean of *all* street addresses, however rather to the mean of the hotspots. We make this distinction because the general rule is that extreme outliers tend to regress closer to the overall mean over time, though this does not mean that they dissipate 'into' the mean completely.

Differential selection

Threats from differential selection are the effects associated with comparing treatment and control groups that are not drawn from the same pool of participants, so the treatment group is inherently different from the control group. The characteristics of participants in the treatment group are not the same as those in the control group. The implication of differential selection is to therefore artificially disadvantage one group of participants before the stimulus is applied, so the test becomes unfair.

This concern is linked directly to selection bias, and for this reason, RCTs have an advantage over other causal designs: randomisation directly deals with selection bias. This does not suggest that experimentalists should not be mindful of the issue of selection bias; randomisation protocols often break down in field experiments, and the risk of confounding is therefore present even in RCTs. In principle, at least, the groups can be assumed to be balanced on known and unknown factors due to randomisation, with nil differential selection effects, but this assumption is often challenged (Berger, 2005a, 2005b, 2006).

For quasi-experimental models, however, *selection bias* remains one of the most fundamental difficulties. Researchers using statistical instruments to control for differential selection go through great lengths to convince the audience that a particular effect is statistically controlled for. However, statistical controls that attempt to account for pre-treatment differences between the groups are unable to remove the *entire* imbalance – in theory as well as in practice. In a population with an infinite number of covariates and their interaction effects, statistical controls can be applied only to a subset of covariates and their interaction effects on which observable data exist. The statistical control over selection bias will therefore asymptotically approach complete control over the effect of the covariates, conditional on the availability of measurable covariate effects. None of this is new: the term *e* for error is part of the formulae of virtually any statistical tests.

Box 3.4

The Error Term

We like the following explanation for the *error* term:

> An error term is a residual variable produced by a statistical or mathematical model, which is created when the model does not fully represent the actual relationship between the independent variables and the dependent variables. As a result of this incomplete relationship, the error term is the amount at which the equation may differ during empirical analysis. (Hayes, 2020)

From a practical perspective, consider studies that evaluate treatments within prison settings. There are several examples of impact evaluations in prison services, which attempt to explore the effect of different treatment programmes on crime after release. One such programme is the religious-oriented rehabilitation programmes on recidivism in prisons. Can a programme based on faith reduce recidivism? We argue that we presently cannot answer this question in full, because you cannot randomly assign religiosity. Controlling for this complicated and internal factor appears impossible (e.g. Haviv et al., 2019), and the evidence remains unconvincing given the selection bias in the data. To study the effect, we need to compare those who joined prison-based programmes with those who did not. But the two groups are fundamentally different. We can match the two groups based on socio-demographic variables or criminal background, but they cannot be matched based on the crucial factor of religiosity. The same can be said about other latent variables such as motivation to refrain from crime in the future or willingness to stay out of trouble. These are unmeasured dimensions that cannot be ruled out using statistical models such as propensity score matching – and further, religion would be a tough dimension to operationalise.

In addition, selection into religious-based programmes in prisons is self-serving and structural: participants volunteer to join these programmes. Volunteers are different from non-volunteers, so the differences between the study groups will be systematic, not random – which is the core of selection bias. Even when trying to control for selection biases by using the participants' level of involvement in the group as a proxy for individual levels of religiosity, programme completion or taking more intensive programmes within the religion-based treatment – the results remain elusive, because we cannot account for the internal processes that drive these measures. For example, we can hypothesise, with reason, that reductions in recidivism after exposure to the religion-based treatment programmes are a function of the therapeutic community environment in which these interventions take place, rather than religion *per se* – which undermines the internal validity of the test. Will we obtain the same results if we mimic the same structural, organisational and social bond aspects of the programme if we replaced the religion component with another profound participatory feature? As we cannot answer this question, selection bias jeopardises our ability to falsify the null hypothesis. To be clear, we are not arguing that faith-based interventions do not work (Harris et al., 1999); we are, however, arguing that we struggle to empirically test their effects because of *differential selection*.

Given these selection bias considerations, we can differentiate between experimental and quasi-experimental designs in the following way: in experimental designs, the researcher is able to control not just the scheduling of data collection procedures (what to measure, when to measure and on whom) but also the scheduling of the experimental stimuli (who will be exposed to the intervention and when). This does not exist in quasi-experimental designs, where the researcher can only control the scheduling of the data collection procedures, not the treatment. Thus, quasi-experimental designs are 'deemed worthy of use where better designs are not feasible' (D.T. Campbell & Stanley, 1963, p. 34).

Experimental mortality (attrition)

When the treatment group experiences greater attrition among respondents from the beginning of the study until the end of the follow-up period than the comparison groups, the internal validity of the test is jeopardised. More generally, *attrition* refers to an issue where participants do not complete the intervention programme. This is commonplace, especially in field settings, when the participants are potentially inclined to complete the treatment – drug users, prolific offenders, unmotivated delinquents and so on. Domestic violence perpetrators are a case in point (Mills et al., 2013; Mills et al., 2019). Attrition in treatment programmes for domestic violence offenders has been an enduring problem ranging from 22% to 99% (Daly & Pelowski, 2000), and, more recently, ranging from 30% to 50% (Babcock et al., 2004, pp. 1028–1030; Gondolf, 2009a, 2009b; Labriola et al., 2008) – which is concerning, because non-compliance in domestic violence treatment programmes remains a strong predictor of reoffending (Heckert & Gondolf, 2005). In our context, it suggests that there may be systematic differences between the completion rates of one group over the other, and this alone may affect the outcome of interest (rather than the intervention itself).

The problem of experimental mortality is exacerbated when it occurs more frequently in one experimental arm than the others. In these cases, the differential attrition occurs *because* of the interaction with belonging to the experimental arm that experiences a differential level of attrition. Therefore, random assignment cannot solve this problem completely – that is, the attrition may be causally linked to the intervention itself. The tested intervention caused participants to drop out for a variety of reasons – for example, the intervention can be too demanding, its participants may feel that the intervention lacks efficacy or the treatment deliverers themselves simply do not 'believe' in the intervention and therefore are unmotivated to deliver it as the experimental protocol dictates. Subsequently, units drop out of one of the study groups at a higher rate than the other study group(s). Those who remain in the programme may have unique features that their overall group does not – for example, more (or less) motivation, more (or less) compliance or more (or less) risk aversion.

One way to deal with attrition issues in RCTs is by using an **intention-to-treat** (ITT) approach. This procedure allows for an assessment of the treatment effectiveness, whether or not it was actually delivered (Gibaldi & Sullivan, 1997). Using ITT, experimenters simply 'ignore' the attrition, or the level of completion of the programme, and focus on the potential effects of offering the treatment *policy* instead of the effects of the treatment-as-delivered. This approach is usually appropriate in field experiments, especially those meant to test treatments that in nature are multifaceted and complex where part of this heterogeneity is the attrition of participants. We cannot force completion of treatments according to an experimental protocol, particularly with experiments involving human subjects, nor can we ignore human error in any human actions. Therefore, an ITT approach provides a solution to the attrition bias by measuring the outcomes of the study of *all* participants *randomly assigned* to the groups, rather than just the outcomes of those who completed the trial (measuring treatment-as-delivered). If we were studying a domestic abuse intervention, for example, in an ITT trial, we would consider the outcomes of *all* assigned participants assigned to the experimental arms. This list would include treatment participants and control participants who failed to turn up, dropped out before the programme was completed or were deported.

Treatment spillover

In causal research, we assume that each participating unit is not affected by any other unit. This means that we assume that any potential outcome that takes place in the experiment is a result of the treatment, not the influence of the other units. If we allocate participants into treatment and control conditions, each group must be independent – otherwise, we may encounter biases in our causal estimates. When the groups are not independent, we experience interference, and we must therefore conclude that the treatment effect was either inflated or deflated, meaning that the true impact of the independent variable on the dependent variable is masked to some degree. This is referred to as a *spillover*, and while it is generally overlooked in criminology (Ariel, Sutherland, & Sherman, 2019), it is one of the most concerning issues in experimental research.

Spillover effects in RCTs contaminate the purity of the experimental design. The diffusion can take many forms, referring to the 'bleeding' from treatment to control or vice versa, between treatment groups, in particular batches, blocks or clusters, and even between individual units within the same experimental arm (Baird et al., 2014; D.T. Campbell et al., 1966; Shadish et al., 2002). For example, when the threat of spillover denotes an interference from the treatment group into the control group, it leads to 'contaminated control conditions', which challenge the 'counterfactual contrast' between units that were exposed to the intervention and units that were not.

More explicitly, the spillover problem was discussed in the context of the 'stable unit treatment value assumption', or **SUTVA** (Rubin, 1980). These are circumstances akin to having both experimental and control patients taking the pill in a drug trial: when everybody is exposed to the treatment effect, the intervention is 'interfering' with the neutrality of the comparison group.

There is another type of SUTVA issue – beyond the interference of one group affecting the other – and that is between the individual units themselves within the same group. When the interference is caused within the group, the treatment spills over from some individuals onto *other* individuals in the *same* group. Sobel (2006) referred to this situation as 'partial interference'. This is an important issue, because causal studies and most statistical models assume that there is a single version of each treatment level applied wholly to each experimental unit, which is referred to as 'treatment homogeneity'. One example would be that every offender assigned to a 12-session domestic violence intervention programme attended all 12 sessions (Mills et al., 2012). Likewise, in a study on the effect of text message nudges sent to offenders as reminders to attend their court hearings, it might be assumed that every individual has received, read and then internalised the messages (Cumberbatch & Barnes, 2018). Similarly, a trial on the effect of omega-3 supplements on behaviour problems in children might posit that the participants adhered fully to the experimental protocol and took precisely 1 g/day of omega-3 during the days of the experiment (Raine et al., 2015).

However, when participants from any of these groups interact with one another, they can increase or reduce the treatment effect.[5] The treatment homogeneity assumption is challenged. For example, if members of a restorative justice experimental arm talk amongst themselves, share their experiences and form a group opinion about the efficacy of the treatment, then these collective experiences and cross-fertilisation with ideas, norms and thoughts violate SUTVA. There is a wealth of ethnographic research that indicates the extent to which group dynamics have an effect on perceptions, behaviours and norms (e.g. Hare, 1976; Lewin, 1948; Shaw, 1981; Thibaut, 2017). Kruskal (1988) commented that nearly all real-life experiments should assume (partial) interference, but ignoring dependence can have detrimental consequences (see applications in Lájer, 2007; and a review in Kenny et al., 2006). When researchers expect 'treatment propagation' (spillover; Bowers et al., 2018; Johnson et al., 2017), they should arguably incorporate it into the experimental design using a group-based model (see Box 3.5 for an illustration).

Box 3.5

Spillover in Body-Worn Cameras Experiments

Related to the choice of the unit lies a critical concern about an old but often neglected issue regarding 'diffusion of treatment' from experimental units to control units, and vice versa. Imagine an experiment on the effect of police body-worn cameras (BWCs) on the rates of documented use of force and civilian complaints against

---

[5]The same rule also applies to the other arms of the test – that is, if there are more treatment conditions, then we assume that each condition was applied fully and equally across units, and that the counterfactual condition (placebos, no-treatment, business as usual intervention etc.) was maintained fully and equally across non-treatment units as well. As you can imagine, this is not always easy.

police officers. In this study, a sample of 2224 police officers randomly assigned into a treatment group were instructed to wear BWCs while on patrol in comparison to the control group, who were not given the devices. In principle, this design is powerful enough to detect small differences between groups. However, there is a catch; the design does not take into account the fact that many – if not most – police–public encounters that require the use of force are dealt with by at least two officers on site. In fact, most police patrols in the contemporary USA are in double formations ('double crewing'). Given these facts, there is a strong degree of 'treatment diffusion' or 'control officers' who attend calls with the 'treatment officers' and are, by definition, 'contaminated' by being exposed to manipulation and subsequently behave differently than when the camera is not present. The risk of *spillover* (another common term for diffusion and contamination) becomes even more pronounced when three or more officers attend the same call. Therefore, the study's treatment fidelity is not only at risk when both arms are exposed to the same intervention.

This spillover explains why such a randomised controlled trial concludes that BWCs are not effective in reducing the rates of complaints or the use of force. It appears that the contamination is so pervasive that an 'intention-to-treat' analysis – that is, one in which all units are analysed in the groups to which they were randomised – would result in no measurable impact (Gravel, 2007). Such a conclusion is unsurprising; after all, both treatment and control officers were treated.

One research design that can most greatly reduce spillover effects is the cluster-randomised trials design, in which entire and remote groups are randomly assigned into treatment and control conditions (see Donner & Klar, 2010). Such designs assume that contamination will occur and incorporate it in the model. It requires numerous groups – entire police forces, entire departments, entire schools and so on. The issue is usually obtaining enough clusters to achieve sufficient statistical power, as power is largely a function of the number of clusters rather than units within clusters. This difficulty explains why cluster-randomised trials are not commonly used in the social sciences, with the notable exception of education. This design could be achieved if a single department operated over a sufficiently large geographical area with enough subdivisions or partners so that it was possible to allocate entire stations, precincts and so on to different conditions. Other options can be used, as forces 'naturally' roll out BWCs that also take into account the clustering.

Additive and interactive effects

Shadish et al. (2012) wrote,

> Validity threats need not operate singly. Several can operate simultaneously. If they do, the net bias depends on the direction and the magnitude of each individual bias plus whether they combine additively or multiplicatively (interactively). (p. 51)

As we have alluded to throughout this chapter, there can be an interaction between the different threats to validity, in which two (or more) threats work together to threaten our conclusion about the effectiveness of the intervention relative to the control group. These are the interaction effects of various factors happening simultaneously – that is, the threat added to that of another threat. For example, we can think of a combination of a history–maturation additive effect, when the treatment participants mature differently than the control participants; however, they are *also* exposed to historical events differentially. Similarly, we can think of an interaction effect between selection bias and instrumentation, where the groups do not share the same population characteristics at baseline (hence the selection bias); however, they are also measured with different instruments (e.g. a team of reviewers – or *raters*, as they are called in research methods – ring fenced for the treatment group and a team of raters ring fenced for the control group). The combination of the two artefacts causes a net threat to the internal validity of the test that goes above and beyond the effect of each of these threats when they operate singly.

Case Study 3.1

Confusion Between Police-Generated and Victim-Generated Crimes in Policing Experiments

We note in this section the possible interaction effect between measurement and outcome. This is problematic in terms of internal validity because we may be confounding the treatment effect with testing effects, or effects caused by the treatment versus those that are caused by the experimenter. This is especially important in policing experiments: a clear distinction between police-generated and witness or victim-generated records is required; however, the police database overlooks this crucial discrepancy. Experimenters need to isolate outcomes from outputs (Ariel, Weinborn, et al., 2016). In policing, outputs are produced through proactive productivity, and outcomes are the result of the outputs. Traffic offences, drugs and tax evasion are all examples of behaviours that are unlikely to be reported by the public (exceptions to these can be found in Ellis & Arieli, 1999; Near & Miceli, 1985). Therefore, increases in confiscations, arrests and detentions for committing these crimes are directly linked to the 'dosage' of law enforcement, not crime per se (Wain & Ariel, 2014). The more attention one of these crime problems receives from the police, the higher the police crime figures will be in that crime category.

On the other hand, crimes that come to the attention of the police through victims and witnesses have a different journey – they are reported to the police independent of a specific police intervention. When victims 'generate' crimes, it is often considered that the trend is not affected by immediate police practices, but rather the willingness of the victim or witness to cooperate with the police (Macbeth & Ariel, 2019).

As recently discussed by Ariel and Bland (2019), policing experiments have differentiated between outputs and outcomes, especially in place-based interventions, when the unit of analysis is the crime hotspot and the test measures police effectiveness (Ariel, Sutherland, et al., 2016a; Sherman & Weisburd, 1995), but it has yet to catch on in epidemiological and longitudinal studies of crime (cf. Sherman, Neyroud, et al., 2016). When this distinction is not applied, it then becomes unclear whether the police's data represent a genuine variation in crime rates or police actions; they are linked, but they are not the same. One alternative explanation is that police tactics and recording strategies dictate fluctuations in some crime figures (Ariel, Bland, et al., 2017; Ariel & Partridge, 2017). When the police target drug or gun offences, these output figures go up (Sherman et al., 1995). For crime categories that are conditional on police outputs – stop and search, arrests, seizures, crackdowns and so on – we would expect an increase in these categories when the police increase attention, because the police have placed these behaviours as performance targets (Martin et al., 2010).

Threats to external validity

When a study is characterised as having internal validity, it is said that there is evidence that the test is able to show that an intervention caused a 'potential outcome' (Fisher, 1935; Kempthorne, 1952; and importantly, Rubin, 1974, 1990b). However, this quality by itself does not mean that the study has *external validity*, which refers to the generalisability of the experimental results. This is a set of concerns about inferences – that is, the extent to which a causal relationship 'holds over variations in persons, settings, treatments, and outcomes' (Shadish et al., 2002, p. 83), and there are different sets of solutions for controlling threats to external validity. We discuss these in this section.

Broadly speaking, maintaining external validity means that we can rely on the outcomes of a particular study – or a series of studies – for policy purposes. For example, how relevant are the experimental findings from the early 1980s about the effect of misdemeanour domestic violence arrests on repeated victimisation (Sherman & Berk, 1984) of domestic violence victims in London in 2020? Are the findings from an experiment on the effects of wearing body-worn cameras (BWCs) in traffic policing in Uruguay (Ariel, Mitchell et al., 2020) pertinent to a police force in Coventry, England, that is trying to increase its perceived legitimacy? Should we roll out a national programme in Australia to prevent dating violence based on an RCT that shows promising results of a Los Angeles school-based dating violence prevention programme among Latino teens from 2006 (Jaycox et al., 2006)? To stress, we are not saying these are *irrelevant* but that the generalisability of experimental results is a continuous *threat* in *any* experiment. As we discuss here as well, the proof for external validity is a measurable replication of the results using the same research design but in different experimental settings.

Box 3.6

Different Designs Yield Different Outcomes

Interestingly, different research designs applied on the same research question yield different study outcomes (see Weisburd et al., 2001). For example, in a meta-analysis on the effect of school-based programmes to reduce bullying, Ttofi and Farrington (2011) have found that only one of the nine randomised controlled trials have shown significant treatment effect, while other experimental designs produced significant treatment effects – leading these reviewers to conclude that, overall, school-based anti-bullying programmes are effective, with bullying decreased by 20% to 23% and victimisation decreased by 17% to 20%.

On a wider level, however, we can consider external validity in three layers (Shadish et al., 2002).[6] These layers will become important as we review the different aspects of external validity:

1   *Narrow to broad*: this is probably the most intuitive and well-known type of concern for external validity. It deals with the question 'Can we generalise from the "persons, settings, treatments and outcomes" of one particular experiment to a larger population?' For example, can we generalise the effect in an experiment on police presence on the London Underground train platforms in 2011–2012 onto all train stations across in England and Wales (Ariel, Sherman, et al., 2020)? In order to be able to reach this conclusion, we must ensure that the platforms and the crimes experienced, the police who have delivered the treatment, the type of treatment delivered and the types of measures the researchers looked at are similar. Otherwise, the transition from the narrow to the broad may be questionable.

2   *Broad to narrow*: here we target a subsection of the experimental sample – for example, in an attempt to check whether the study findings are relevant to a *single* person or a group of individuals, one could ask whether restorative justice conferences that were found to reduce *overall* post-traumatic stress symptoms in victims compared to similar victims who did not go through this intervention (Angel et al., 2014) would work for one particular crime victim. If the treatment effect was totally ubiquitous (like gravity), then we can see how – and why – the generalisation from a broad finding that the intervention is effective in reducing post-traumatic stress symptoms in any one individual victim. However, in the social sciences, we rarely ever have such strong interventions, which is why external validity always remains a concern, especially in broad to narrow generalisations.

---

[6]Shadish et al. (2002) in fact discuss five targets. However, we feel that the remaining two are somewhat confusing and do not discuss them here.

3   *At a similar level*: here we consider whether the conclusions are transferable to similar samples that share the same aggregation (e.g. similar settings). For example, whether or not we can generalise from one cluster RCT with 36 schools in London, intended to reduce school exclusions through a 12-week-long intervention – Engage in Education, London, delivered by Catch22 (Obsuth, Sutherland, et al., 2017) – to Manchester or Birmingham settings. In this context, the question of external validity is particularly important, as the experiment produced mostly null and even backfiring effects.

Under the bonnet of threats to external validity

There are always risks to the generalisability of the results of any study (D.T. Campbell & Stanley, 1966). Our ability to say that the findings are relevant to people outside the sample on which the test was conducted, or in terms of other places and settings, or to withstand the test of time, is, by definition, limited. There are both theoretical and methodological underpinnings to this issue of generalisability.

First, scholars prefer theories that withstand the test of time, location and context – however, they will immediately admit that this aspiration is largely conjectural in the social sciences. No one variable or set of variables has the capacity to fully explain the totality of causal mechanisms of crime, deviancy or human behaviour more broadly (Hedström et al., 1998). Therefore, we should not expect the findings from any one experiment to explain all variations in the universe, even if conducted meticulously and flawlessly, because there are no univariate relationships between dimensions. There is no *one* independent variable that affects *one* dependent variable, for all participants, all places and all contexts, and by implication just one theory that fits to explain this relationship (Young, 1980). Thus, while we might be able to prove a cause-and-effect relationship, we are not always able to prove that the relationship is replicable in settings that are slightly different from those in which the relationship was observed.

In comparison, take gravity: for ordinary purposes, the earth exerts the same gravitational pull on people who are close to its surface (9.807 m/s^2), and in the same direction (down). Gravity is ubiquitous and works on all people, every time and in any context (Minstrell, 1982). Any experiment that would include gravitational pull as a causal factor would repeatedly yield the same result – thus maintaining high levels of external validity. In the social sciences, however, we rarely encounter effects that are as strong and as consistent as gravity, and we should always expect variations when attempting to replicate experiments; culture, norms, attitudes, aptitudes and behaviours are complex and multifaceted concepts, and they tend to be time, place and person dependent – which, collectively, threaten external validity. Thus, while the law of gravity is strong, our theories in criminology, sociology, psychology, epidemiology, economics or education are not as sturdy.

The second important factor is directly related to the first, but on a more technical level. Indeed, the assumption *is* that we cannot achieve 100% reproducibility in the social sciences, and we are unlikely to repeat the same results with the *exact* same direction, statistical significance level and effect size – and realistically we should expect some variations from the findings of any original experiment. However, how can we *aim* to achieve *as much* replicability of findings as possible? This is a methodological concern: implementing guidelines for creating the necessary settings in which the experimental settings can be *reasonably* repopulated. For example, any social science experiment conducted in entirely bespoke settings is unlikely to carry high external validity, because these settings cannot be repeated outside the scope of the study. Similarly, if the participants are unique to the point that no other persons outside the study parameters resemble them, the experiment is likely to suffer from low external validity as well. Therefore, we need technical guidelines on how to set up experiments that have a reasonable level of external validity; otherwise, the outcomes will not matter much outside the specific environment of the original study. We will provide some of these guidelines in Chapter 4. First, though, let us examine in more specifics the major threats to external validity in terms of participants, places, settings and time.

Participants

The very first threat that experimenters must deal with is the issue of selecting – intentionally or unintentionally – units that are inherently different from the population from which these units were sampled (R.M. Martin & Marcuse, 1958; Rosenthal, 1965). One of the most important examples is the use of volunteers in field experiments: participants who *choose* to take part in a study on a charitable basis may be unrepresentative of the population from which they were recruited. They may, for example, be altruists, a quality that makes them different from the population to which they belong. The problem is exacerbated when the unit of analysis is an entire volunteering organisation: the willingness of an entire police force, an entire school or a whole treatment facility to participate in an experiment may be an indication of difference from all other organisations (Levitt & List, 2007b).

Similarly, those who participate for cash incentives may be different from those who are uninterested. The same types of volunteers may not be available in other settings – and in fact may not be available at all in real-life settings once the experimenters have left the research site. Take, for example, experiments that use online platforms to recruit participants (e.g. Amazon MTurk, SurveyMonkey): they suffer from misrepresentation,[7] and while they are appealing to researchers, the extent to which certain people are hired to participate in a clinical experiment online are similar to the kinds of populations criminology is interested in remains an unanswered question. In policing, we also have a similar issue when it comes to volunteers. There is nothing methodologically wrong about using volunteers, for-pay participants or enthusiastic do-gooders – however, the researcher must hedge the conclusions derived from these specific populations – as opposed to the overall population from which these volunteers were recruited (Jennings et al., 2015; Ready & Young, 2015).

The concern with external validity in terms of people is not just about the willingness of the participants to take part in an experiment. When the sample under investigation is unique, or when the cohort that takes part in the study has distinct features, we may not be able to conclude that the findings are transferable to other samples or cohorts that do not share these features. Can we generalise the findings from an experiment conducted on Danish gang members to non-Nordic gang members (Højlund & Ariel, 2019)? Are lessons learned about gang injunctions in Merseyside, UK, relevant to US gang problem (Carr et al., 2017)? To what extent are prisons in Israel similar to prisons in England and Wales (Hasisi et al., 2016)? Will the same conclusions found on the mass deployment of Tasers in one small force in London be found in larger forces outside of London (Ariel, Lawes et al., 2019)? If there are social, cultural or background differences between the original experiment and those of the target population, then we risk making errors in the translation of the conclusions to the target population(s). To emphasise, we are not suggesting that these experiments are *not* translatable, but simply that we need evidence to support the hypothesis of generalisability.

Finally, we note that the issue of external validity is more profound when only one group – for example, the treatment group – is required to participate in the study, but not participants from other groups. Not only will the groups be unbalanced in their willingness to participate, a source of concern in terms of internal validity, but we may also find it difficult to generalise to the overall population if the treatment group participants must express their consent to participate *after* random allocation. Under these conditions, the treatment group is 'better off' than the control group, as those who are willing to take part in the treatment are often more motivated, engaged and have a better prognosis to succeed than the control group, which is made up of such individuals *as well* as individuals who are unmotivated, disengaged and in poorer conditions.

---

[7] We question the generalisability of *all* experiments which use these volunteers or for-pay participants because they may have different qualities than those who do not go on these websites to search for studies in which they can participate. While we see the merit in these studies, especially in terms of costs and convenience, we also see external validity concerns that are difficult to control – particularly about how representative the population of participants is of the overall population (not least by way of access to the internet, language barriers etc.).

Case Study 3.2

'Creaming' of Participants as a Threat to External Validity in Pennsylvania

'Creaming' refers to the non-random selection of participants into one of the experimental arms who are the most likely to react positively to the treatment, but in a way that creates a selection bias. In other words, the treatment provider selects participants who are more likely to achieve the expected outcomes but leaves outside the participants with the most challenging cases. An example of this concern came up in the Bethlehem Pennsylvania Police Family Group Conferencing Project (McCold & Wachtel, 1998). In this experiment, juvenile offenders were randomly assigned either to a diversionary 'restorative policing' process called family group conferencing or to control conditions. Under this scheme, police officers facilitated a meeting attended by juvenile offenders, their victims and their respective family and friends to discuss the harm caused by the juvenile offender's actions. The goal was to develop a restorative justice agreement to repair the harm. Participation in this scheme was voluntary – by both the victims and the offenders. Compared to the formal adjudication group (control conditions), the study suggested benefits to victims and offenders, based on a series of surveys (on methodological issues; see Strang & Sherman, 2012).

However, the study design was such that only *after* a case was assigned to the treatment group did a police officer attempt to elicit participation from the offender and then the victim. Only

> where both offender and victim were willing to participate was the case assigned to the facilitating officer. If either party was unwilling to participate, the case was not conferenced and, thus, was processed through normal channels like the control cases. (McCold & Wachtel, 1998, p. 17)

Thus, those who eventually participated in the conferences were qualitatively different from those who dropped out in the treatment group *and* from those who were randomly assigned into control conditions. Dropping out was not an option for participants in the control group. This means that the final two study groups – adjudication-only group and volunteers in the restorative justice group – were inherently different even before study commenced. The study group participants were inadvertently creamed and, by definition, better off than the population of juvenile delinquents from which the unit was drawn (not least in comparison to the comparison group, which creates systematic variations that lead to self-selection issues) – thus making any conclusion about the efficacy and cost-effectiveness of the intervention weak.

Place and settings

How transferable is knowledge gained under some settings to others? It is not easy to answer this question. Consider, for example, focused deterrence studies in the USA (see review in Braga, Weisburd, et al., 2019). In a series of experiments, which date back to the 'Pulling Levers' project outlined in Boston ('Operation Ceasefire'; Braga et al., 2001), police officers and community leaders worked together to reduce gang-related violence. The core police tactic was to increase the certainty, swiftness and severity of punishment in a number of innovative ways, often by directly interacting with offenders and communicating clear incentives for compliance with and consequences for criminal activity.

However, are the settings that characterise urban US gangs relevant to rural UK gangs? Is the level of resourcing obtained by an expensive programme like 'Pulling Levers' replicable to Coventry anti-gang units (see discussion in Delaney, 2006)? Is the level of harm perpetuated through firearms in American cities similar to the hand-to-hand combat and knife crime that characterises the majority of urban street gangs in the UK? Is the level of funding for research similar to the amount of money available for scholars in the UK? The experimental settings in the 'Pulling Levers' projects may not be immediately transferable to other places in the USA or to other countries – and maybe even to the same locations where these projects were implemented, however, in later years.

Time

One of the major difficulties in generalising from a particular experiment is time. How much can we conclude from a study that was conducted in the 1950s, for example, to the 'temporal settings' of 2020?

Take the famous 'Connecticut crackdown on speeding' experiment (D.T. Campbell, 1968). In 1955, Connecticut experienced a heavy death toll in highway traffic accidents. Speeding was suggested as the major cause of this phenomenon, so Governor Abraham Ribicoff increased sanctions against speeding by suspending the licences of drivers for 30 days, 60 days or lifetime suspensions, depending on the number of times they were apprehended. After implementing the new initiative in 1956, the results were encouraging for the first six months: a 15% reduction in deaths compared to the same period in 1955. While there are fundamental flaws inherent in this conclusion, let us assume that we can take the conclusion of the original analysis at face value: that speeding can be causally associated with fatal accidents and that the enforcement of traffic violation through tickets has a causal effect on speeding.[8] However, this study was conducted more than half a century ago. There had been substantial improvements in road safety systems, safety features of vehicles, medical technology and information technology. Speed cameras, airbags, seat belts and artificial

---

[8]More recently, Luca (2015) has argued that tickets significantly reduce accidents but that there is limited evidence that tickets lead to fewer fatalities, and a meta-analysis synthesising the evidence concluded that the state of the art of the research is generally weak, suggesting that 'estimates of changes in violations or accidents should be treated as provisional and do not necessarily reflect causal relationships' (Elvik, 2016, p. 202).

intelligence sensors are only a few elements that make the original Connecticut study somewhat obsolete (Ariel, 2019; Høye, 2010, 2014; Phillips et al., 2011). It is also the case that the driving culture has shifted since, not least in terms of drunk driving (see Jacobs, 1989). Therefore, the tremendous leap we have made in road safety casts doubts on the generalisability of the original study to the present day.

Increasing the external validity of a test

There are two major ways to directly address threats to the external validity of a test: **replications** and **random sampling**; while the former is well accepted and ordinary, the latter is less attainable in the social sciences. External validity could be increased by conducting the same research on more occasions across different populations and settings. Diversifying the portfolio of research on the *same* research question, however, in other settings, with different participants and over time, reduces the risk of a false finding obtained from the original study. Replications also allow us to understand not only whether the original conclusions can be generalised to different target environments but also to what extent. The more experiments in these varying settings, the more we can understand on whom, and under what conditions, the treatment in question can 'work' (Lösel, 2018; cf. Gilbert et al., 2016). Therefore, the best way of increasing the external validity of experimental findings is replication.

Second, external validity can be greatly enhanced through proper sampling techniques. By using 'probability sampling', the sample with which the study is conducted can then represent the population from which it was drawn. In probability theory, probability sampling is a technique where the sample of participants is *randomly* drawn from the population – which, by definition, has the best chance of representing that population. There is a great deal of methodological literature in this area (e.g. Fitzgerald & Cox, 1994); however, the crux of the argument is relatively straightforward: through a process of random selection, we have a mathematical method of assuring sample representativeness, some of which are discussed below.

In order to ensure the representativeness of a study sample, three things must occur: first, that every unit has a known probability to be selected from the sampling frame. This means that there must be a list from which we can *randomly* select units to participate in the experiment. Note that random *selection* is not the same as random *assignment*; the latter refers to the random assignment of units into the study groups, whereas the former refers to the process of obtaining units from the population. Thus, we select a finite number of units from the population randomly, and each unit has a known and predictable chance of entering the experiment. In practical terms, if we decided that the experiment will include 100 participants out of a population of 10,000, then each participant has a 1% chance of being selected into the study – just like everybody else. Then, by either using a simple RAND function in Microsoft Excel or any other computer software, we can randomly choose the 100 units and exclude the other 9900 (McCullough & Wilson, 2005).

The second and third rules about random sampling are that there is no unit in the sampling frame that has a guaranteed chance to be included in the study (rule 1), or a guaranteed chance *not* to be included in the study (rule 2). These rules ensure that we will not have a unique set of characteristics in the sample vis-à-vis the population, a scenario that reduces the generalisability of the conclusions.

Despite these acknowledged rules (see Lohr, 2019), experiments based on true probability samples are rare, and the truth is that most police experiments are based on convenience or purposeful samples. These are sampling techniques in which the experimenter selects units based on their availability for practical reasons (e.g. treating all hotspots that have a certain threshold of crime levels) or statistical terms (e.g. achieving statistical power – see explanation and expansion in Britt & Weisburd, 2010, as well as a primer by Jacob Cohen, 2013). Any experiment in which all known hotspots in a city above a certain threshold of 'heat' are selected to participate in the experiment (see Duckett & Griffiths, 2016), all domestic offenders in a certain area are placed in either treatment or control conditions (Strang et al., 2017) or all eligible victims and witnesses in a particular force are chosen to receive text messages to mobile phones inviting them to appear in court to reduce their non-appearance rates (Cumberbatch & Barnes, 2018), is *not* a probability sample. This means that, by definition, we should not expect the units to represent all places, domestic violence offenders or victims and witnesses who are scheduled to appear in court. Thus, real-life RCTs are often conducted with a specific population or problem in mind, so the issue of external validity, in the statistical sense, is unanswered.

An exception to this is experiments in which there are more eligible participants in the population from which the units are drawn than available treatments, and then the researcher can implement probability sampling. For example, drug treatment facilities that have the capacity to treat n participants at any given time; however, the geographic region may have a larger number of drug addicts who can be assigned to treatment and control conditions. Another example is tax compliance research, where the entire population of taxpayers are potentially eligible to participate in an experiment; however, the tested intervention can only be applied on a limited number of taxpayers (e.g. Ariel, 2012). These studies, however, seem to be rare in criminology given the types of populations with which the police normally interact.

Case Study 3.3

Hawthorne and John Henry Effects

In experiments, we must accept the possibility that the observed effect may have followed from a *Hawthorne effect* or a *John Henry effect*.

Hawthorne effect: These are situations in which participants change their behaviour simply because they are taking part in an experiment. The name comes from a famous study in which this effect was first illustrated. As reviewed by H.M. Parsons (1974), seven studies took place between 1924 and 1932 and grew to be associated with this effect. These tests took place at a Chicago plant where the Western Electric Company produced various types of equipment for the Bell Telephone System (Mayo, 2009). The experiments were aimed at increasing the workers' productivity, particularly in the Relay Assembly Test Room. Female participants performed product-relaying tasks that required procedural memory and visual discrimination, manual dexterity and coordination. To assemble a product, a worker would take parts out from boxes, reject faulty parts and arrange the parts together into a final product. The assembly was meticulously observed (thus being able to show e.g. that 32 motions of the right hand and 31 of the left were employed for each relay product assembled). Over the course of the experiment, each operator assembled 100 to 150 relays (Landsberger, 1958; Mayo, 2009; Whitehead, 1938).

The experimenters tried to increase productivity in the Relay Assembly Test Room by manipulating different things. The electric company asked the research team to determine if there was a link between work environments and productivity, including how the relays were assembled, the timing of breaks or varying the illumination (the latter manipulation arguably being the most famous). Elton Mayo and his colleagues changed the ways in which the workers performed their jobs in the Western Electric Company. However, it was shown that, no matter what changes were applied by the researchers, participants worked progressively faster. As a result, the researchers reported that *any* change to the study conditions led to increases in productivity.

The conclusion was crucial for the social sciences: productivity was increased because the workers were observed by the research team. The alteration of behaviour by the study participants was related to their awareness of being observed by the supervisors, or the attention they received from researchers, rather than because of any manipulation of independent variables.

Admittedly, later research has found that the original reports were somewhat misleading. Levitt and List (2011) reanalysed the data from the original Hawthorne studies and found that the conclusions were overstated by the available data. The overall finding is that a 'Hawthorne effect' did exist in the factory; however, it was of smaller magnitude than originally reported. Still, the concept itself remains relevant because it alerts experimenters to the possible biases in generalising the treatment effect beyond the experimental context.

John Henry effect. This 'compensatory rivalry' effect (Heinich, 1970) is similar to the Hawthorne effect, but it differs in that it concerns the behaviour of the control rather than treatment group participants. It refers to situations where members of the control group are cognisant of their status as members of the control group, and when they compare their performance with that of the treatment group, they attempt to overcome the 'disadvantage' of being in the control group by performing better than they normally would.

The origin of the term was explained by Salkind (2010):

In the 'Ballad [of John Henry]', title character John Henry works as a rail driver whose occupation involves hammering spikes and drill bits into railroad ties to lay new tracks. John Henry's occupation is threatened by the invention of the steam drill, a machine designed to do the same job in less time. The 'Ballad of John Henry' describes an evening competition in which Henry competes with the steam drill one on one and defeats it by laying more track. Henry's effort to outperform the steam drill causes a misleading result, however, because although he did in fact win the competition, his overexertion causes his death the next day (Salkind, 2010).

Overcoming the threats to external validity through replications and diversification

The larger question we must answer is the extent to which the experimental conditions mirror real-life conditions. When the difference between the two is 'too large', then the results of the experiment may be deemed unacceptable for policy purposes, or at the very least ungeneralisable. What is most disheartening, however, is that these issues are not logically induced as they are with internal validity: experimenters have to make assumptions about the representativeness of the experiment and the distribution of the data, given the experimental conditions.

In a purist sense, there is no way to generalise from one experiment. It is the accumulation of evidence, over time and across diversified settings, that set up the logical conditions for generalising from experiments. Even the most powerful experiment remains a single unit of knowledge, and the degree to which we can learn from one data point with regard to other settings is always restricted. A thousand experiments with the same research question and different conditions that produce a consistent outcome will always carry more weight through greater external validity. Whether 70 experiments on hotspots policing (Braga, Turchan, et al., 2019), 20 experiments on police legitimacy (Mazerolle, Bennett, et al., 2013; see also Sargeant et al., 2016; Walters & Bolger, 2019) or a dozen experiments on BWCs (Lum et al., 2020) is a 'sufficient' number of trials for laws governing the causal link between these interventions and certain outcomes is contestable. The more we accumulate evidence, the more likely we are to reach this point – in other words, we need replications in field settings.

Distinctions between laboratory and field trials

From our discussion of external validity, it should be immediately clear why laboratory experiments suffer greatly from external validity threats, and why field experiments (by which we mean those trials which take place in natural, live or operational settings – e.g. a school, a police beat or a hospital) aim to overcome these issues. Levitt and List (2007a) argued that we could find it difficult to interpret data from lab experiments, as they are not immediately generalisable to the real world. The authors contend that laboratory findings fail to generalise to real-life settings. Study participants behave differently when they are in the sterile conditions of a laboratory, when often the observed behaviour is hypothetical, rather than real, and when field settings do not naturally confound the independent or the dependent variables.

To explicate the external validity issue of lab experiments, take, for example, those that aim to understand how Taser stun guns affect police officers' use-of-force decisions (e.g. Sousa et al., 2010). This is an important area of study, not least because Tasers remain a contentious tactical option in policing. At the same time, these studies are rarely true field experiments and should be viewed as laboratory experiments – even though the participants are actual law enforcement agents – because they ask how officers behave 'as they *would* in a natural setting' (Sousa et al., 2010, p. 42), not how they actually *do* behave in the field. Instead, scholars tested training scenarios involving different levels of suspect resistance, with police trainers performing the roles of suspects. The 'suspects' are police officers and the decision to 'use' force does not happen in the stressful settings of the field. Therefore, the experimental settings do not fully mimic natural settings – and in fact, they ought not to be judged as such. These lab experiments are crucial to understanding what *might* possibly happen in police–public contacts; however, we are nevertheless unclear as to what extent they reflect what *does* happen to cops who are confronted with resisting suspects.

This is not to say that laboratory experiments are not important. As remarked by D.T. Campbell and Stanley (1963), 'an ivory tower artificial laboratory science is a valuable achievement even if unrepresentative, and artificiality may often be essential to the analytic separation of variables fundamental to the achievements of many sciences' (p. 18). Thus, there are merits to laboratory experimentation insofar as they help lay out future hypotheses and to construe the necessary dimensions of theoretical developments.

Still, we think that Levitt and List's (2007a) critique remains valid (cf. Kessler & Vesterlund, 2015), for the reasons we highlight in this section. Stimuli used in lab experiments do not fully resemble the stimuli of interest in the real world, the lab participants do not resemble the individuals who are ordinarily confronted with these stimuli and the context within which actors operate do not resemble the context of real-life interests (Gerber & Green, 2011). Subsequently, for this (and other) reasons, field experimentation directly attempts to simulate, as closely as is possible, the conditions under which a causal process occurs, with the aim to enhance the external validity of experimental findings (Boruch, Snyder, et al., 2000).

The essential differentiation between the two experimental settings – laboratory versus field environments – is the extent to which field experiments aim to mimic the natural surroundings of the participants. For impact evaluations, field research is optimal, as it provides the strongest case for generalisation (Farrington, 2006). However, even under these natural conditions of real-life settings, there may still be perils to the external validity of the test – as we have tried to illustrate in this section.

With that being said, we must also take into account the laboratorial narrative, which may be suspicious of field tests given the lack of experimental control with which field conditions are usually characterised. Clinical trials, psychological experimenters or biologists are more than happy to sacrifice external validity in favour of internal validity, which in its purist form necessitates clinical settings. To control for all exogenous factors implies shutting down any interaction between the intervention and the externalities – an impossible task for field trials but a possibility for laboratory trials. To single out a treatment effect, in a total way, is an option that cannot be materialised in real-life settings. Thus, to surgically probe a hypothesis (Popper, 2005) entails sterile laboratory conditions. The degree to which these findings are applicable to non-laboratory conditions then becomes a contestable question, and it is often impossible to move these field settings into the laboratory; but there is no doubt that the most stringent levels of controls can be applied in closed settings.

One final point about the connection between treatment integrity and external validity

Directly linked to the question of SUTVA mentioned earlier is the need to apply the treatment identically on all participants and without interference between the units. As the discussion on treatment spillover shows, in studies with human beings one should expect the 'human condition' to get in the way, and researchers should do as much as they can to implement the treatment as intended (Perepletchikova, 2011). Thus, 'treatment integrity', also known as 'treatment fidelity', is integral for empirical testing of intervention effectiveness as it allows for less ambiguity when interpreting the obtained results.

However, it is not easy to apply the treatment uniformly and thus maintain 'treatment integrity' (Perepletchikova & Kazdin, 2005; Sanetti & Kratochwill, 2009). For example, attrition is plausible in a discipline such as criminology, where participants often lead chaotic lives: law enforcement officers work in time-sensitive environments where the precise application of treatments across all cases is difficult. There are several examples: spending the same number of minutes – researchers are rarely dedicated to only one study site and there will be situational, professional and substantive reasons – legitimate and otherwise – where the experimental protocol will be breached. In this sense, spillover effects – especially in the administration of a single value and version of the treatment – can be unavoidable. They are especially inescapable when the sample size is large and precise administration across all units is more challenging (Weisburd et al., 1993).

For example, some participants will take up their allocated treatment, such as therapy or 'pathway treatment', as ascribed by the treatment provider, while others will take part only partially. Likewise, some hotspots may be visited by the police as assigned by the experimental protocol – for example, 15-minute visits, three times a day – nonetheless, other hotspots will be patrolled to a lesser degree. In both these examples, the overall treatment effect may lead to statistically significant differences between the study arms; however, the effect size may be diluted. This was the case in several experiments testing the application of technological innovations in policing (see review in Ariel, 2019) regarding hotspots policing studies, and in batterers intervention programmes, to name a few.

Conclusion

Our aim in this chapter was to cover the common threats to the internal and external validity of the test. Insofar as internal validity concerns are raised, it is vital that the risks to the claim of causal inference will be mitigated as much as possible, preferably through design but also using statistical models, when permissible. When it comes to external validity, there is no true statistical solution, because the degree to which one trial's results are transferable to other settings is ad hoc. The burden of proof of generalisability is empirical, not logical, and appears through replications.

We paid attention to the distinction between true experiments and quasi-experimental models. However, we stress that randomisation alone does not solve all internal validity concerns (Fitzgerald & Cox, 1994), and it provides no solution to external validity concerns. While these threats are therefore more pronounced in quasi-experimental designs, close attention should be given to these threats in RCTs as well. As we explained, there are at least two reasons for that. First, not all randomised experiments are created equal.[9] There are different types. Some designs, like the pre-test–post-test control group design, are more rigorous than others, such as the post-only control group design; the former

---

[9]For a comparison of different experimental designs, see, for example, the Maryland Scale (Farrington et al., 2002; Sherman et al., 1998) and other metrics (e.g. Hadorn et al., 1996).

have a greater capacity to eliminate the threats to internal validity. Some designs, like the Solomon four-group design, are so rigorous that they are almost un-implementable in field settings (or at least we rarely see them in use in the social sciences). Since there are different experiments, there are also different levels of each threat, and we cover these in Chapter 4.

Second, the very assumption of equilibrium due to randomisation depends on a number of postulations – that is, the necessary conditions covered in Chapter 2 needed for the effect of randomisation to take place. These are not easy to assemble. Even in laboratory conditions, sterility often breaks: dirty test tubes, heating/air conditioning malfunctions or uncooperative undergraduate students who take part in experiments for university credit. Randomisation alone thus cannot remove these threats. One clear reason is that the threats do not occur in silos: they interact amongst each other and create new concerns that, again, random assignment cannot eliminate. Therefore, threats to internal validity remain a concern in any trial.

Still, we need to think about the threats to internal validity in a contextual way – that is, in relation to *other* experimental designs. There are many different types of experimental designs, and some are better equipped to deal with internal validity then others. As we discussed in Chapter 2, the random allocation of units into different groups deals directly with issues of internal validity and minimises them the most *in comparison* with other experimental designs. Randomisation – *if* conducted properly – reduces the likelihood that our conclusion about the causal relationship between the intervention and the outcome, relative to the control group, is erroneous. But within the world of RCTs, there are different designs, and each is a better fit for the experimental settings under investigation. Another type of threat to the validity of the test is external validity.

External validity is multifaceted and complicated and stands at the heart of criticism against experimental designs. To what extent can we generalise from a single experiment to other settings, people, places or times? This is a tough question, because it can only be illustrated ad hoc, not *a priori* – unless the sampling frame from which units were randomly assigned is taken from the same population of units. Access to pure random samples of this sort is difficult to achieve in the social sciences. Thus, total external validity is aspirational. There are ways to mitigate it somewhat, like running field experiments in lieu of clinical trials, or having sufficiently large probability samples, but the primary method of ascertaining generalisation at different levels is to replicate the test, in diversified settings with different experimental designs but achieving the same statistical result. The story of different experimental designs is told in the next chapter.

Chapter Summary

- In this chapter, we take a closer look at the *controlled* aspect of RCTs, including the purpose of establishing counterfactuals, the supervision of experimental conditions and practical solutions to managing trials in the field.
- The chapter provides a rigorous treatment of the theory and practice of establishing internal and external validity, the potential threats to these and how controls can wholly or partially remove these threats.
- We show how the need for control can be materialised by applying a set of guidelines as to what experiments should consider.

Further Reading

Shadish, W. R., Cook, T. D., & Campbell, D. T. (2002). *Experimental and quasi-experimental designs for generalized causal inference.* Houghton Mifflin.

Shadish et al. (2002) are often credited with providing one of the most detailed explanations of the various threats to the validity of the conclusions of a study. While this chapter discussed internal and external validity, many other hazards to validity must be guarded against. For a more comprehensive exposition of these threats, as well as other validity concerns in experimental and quasi-experimental designs, this seminal book should be consulted.

4

T IS FOR TRIALS (OR TESTS)

Chapter Overview

Different kinds of trials

In a seminal text, Campbell and Stanley (1963) lay out a taxonomy of experimental design options for the social sciences. There are many such designs – in terms of their ability to overcome threats to internal validity and external validity (Campbell, 1957), as well as their applicability given the real-life conditions experimenters must face. Campbell and Stanley (1963), Shadish et al. (2002), Cook and Campbell (1979), as well as most accepted texts on quantitative research methods in various fields (Bernard, 2017; Cox, 1958; Davies & Francis, 2018; Dawson, 1997; Fitzgerald & Cox, 1994; Jupp, 2012), categorise experiments into three types: (1) pre-experimental designs, (2) true experimental designs and (3) quasi-experimental designs. Within each type, several options for causal research are possible, depending on the research questions and the availability of experimental settings. This breakdown is useful, because it sorts experimental designs based on their ability to control for internal validity threats. Scholars can then broadly agree on the strength of evidence, based on which policy recommendations can emerge. Of course, the ability of the researcher to execute a trial that deals with all threats to its validity concerns defines the quality of any particular piece of evidence. However, at a conceptual level, ranking studies using this logical framework is helpful – and we will follow the same tradition.

Figure 4.1 summarises these designs, while this chapter presents these experimental methods more elaborately. We rely greatly on the original taxonomy offered by Campbell and Stanley (1963) but provide a selective ontology of contemporary experimental criminology to illustrate the designs. Altogether, we describe 13 experimental designs – three pre-experimental designs, three true experimental designs and seven quasi-experimental designs.

As we describe in greater detail later in this chapter, experiments can be arranged on a scale, ranked by the degree of control the researcher has over the administration of the test. *True experiments* incorporate the random assignment of participants into the treatment or control groups of the study, as well as reliable measures before and after the administration of the intervention. In these prospective designs, the researcher is actively – and therefore prospectively – involved in the systematic exposure of treatment participants to the stimulus. *Pre-experimental designs* are also prospective, and the researcher has a degree of control over the implementation of the treatment(s), except that they omit one or more features of the true experimental design. For example, they often do not include a pre-treatment measure prior to the exposure to the manipulation, or lack a parallel comparison group, or randomisation in the allocation of the participants into the experimental arms. On the other hand, in *quasi* (i.e. resembling) experimental research designs the researcher is usually not explicitly involved in the assignment of participants into

Pre-Experimental Designs

X = Observation of Treatment

O = Observation of No Treatment

R = Randomisation

1. The One-Shot Case Study

    Group 1: X O

2. The One-Group Pre-test–Post-Test Design

    Group 1: O_1 X O_2

3. The Static-Group Comparison Design

    Group 1: X O_1

              O_2

True Experimental Designs

4. The Pre-test–Post-Test Control Group Design

    Group 1: R O_1 X O_2

    Group 2: R O_3 O_4

5. The Post-Test-Only Control Group Design

    Group 1: R X O_1

    Group 2: R O_2

6. The Solomon Four-Group Design

    Group 1: R O_1 X O_2

    Group 2: R O_3 O_4

    Group 3: R X O_5

    Group 4: R O_6

Quasi-Experimental Designs

7. The (Interrupted) Time-Series Experiment

    Group 1: O_1 O_2 O_3 O_4 X O_5 O_6 O_7 O_8

8. The Multiple Time-Series Design

    Group 1: O_1 O_2 O_3 O_4 X O_5 O_6 O_7 O_8

    Group 2: O_9 O_{10} O_{11} O_{12} O_{13} O_{14} O_{15} O_{16}

9. The Equivalent Time-Samples Design

    Group 1: X O O X O O

10. The Non-Equivalent Control Group Design

    Group 1: O_1 X O_2

    Group 2: O_3 O_4

11. The Separate-Sample Pre-test–Post-Test Design

    Group 1: R O_1 (X)

    Group 2: R X O_2

    - - - means comparison of parallel groups without random assignment

Figure 4.1 A taxonomy of experimental designs

Note. Adapted from Campbell and Stanley (1963).

the have control over the exposure of the participants to the treatment, because the intervention usually had already occurred *before* the experimenter became involved in the study. Instead of random allocation, quasi-experimentalists aim to control for the lack of comparability between the groups using statistical models and matching techniques.

Of the three research designs, quasi-experimental models are the most prevalent in the social sciences, followed by pre-experimental designs, and then true experiments. However, our intent is to push experimenters to conduct true RCTs and use other modalities only when true experimental designs are not feasible, for whatever reason. The choice of design will depend on practical considerations, the ability of the researcher to control for issues that may affect the validity of the test and what options are obtainable, out of the different methodological scenarios the experimenter encounters. Our underlying position, however, is that true RCTs are inherently stronger and provide the most valid causal estimates, of all research designs. Any other design is, paradigmatically, a compromise, for the reasons we discussed in Chapter 2 concerning the benefits of randomisation and the issues associated with statistical matching.

That being said, we do not *prima facie* dismiss evidence gathered by way of other designs. While some evidence can be considered 'better' than others (Sherman et al., 1998), we nevertheless take the view that scientific exploration is founded on cumulative wisdom, and there are useful takeaways from every study. We would not endorse a general rule based on the outcomes of a single experiment, even a trial that is based on the most rigorous experimental conditions – as internal validity is part of a wider concern regarding the strength of evidence (like external validity). No single trial should be deemed authoritative (Lanovaz & Rapp, 2016). The accretion of multiple blocks of evidence shapes our knowledge about causal relationships between independent variables, dependent variables and factors that shape their interactions (Schmidt, 1992). We discuss these points as we go through the various experimental designs in this chapter.

We also present common methodological approaches to analysing the results of these designs, in terms of statistical tests, but in broad terms only. One frustrating problem in experimental designs is the inability of scholars, especially technicians, to agree on 'best practices' for analysing the results of a given experiment. It is likely that a 'best practice' proposition is a fallible suggestion; experimenters know that there are many ways of estimating the causal relations using statistics, and statisticians cannot always agree with what is the most fitting statistical architecture for each experimental design. As the old British expression goes, 'there are more ways of killing a cat than choking it with cream', and in the case of statistics, this is especially true. As we will show, the statistical instruments available to quasi-experimenters are even more complex, so we will only briefly introduce how they are used to measure treatment effects. The technical literature should be consulted for a more elaborate exposition of these statistical tools (e.g. Altman, 1990; Cohen, 2013; Spiegelhalter, 2019; J.P. Stevens, 2012; Weisburd & Britt, 2014). First, however, we will begin by describing pre-experimental designs.

Pre-experimental designs

Design 1: The one-shot case study

This design incorporates a single group, which is measured only once after the administration of the treatment. Here, a selection of participants is exposed to the stimulus, after which their response to the treatment is observed – but without any measure of their responses at baseline (i.e. at pretest level), or without any comparison. As you may recall from Chapter 1, the lack of a benchmark is cataclysmic for the causal inference claim (see discussion in Berk, 1988). As Campbell and Stanley (1963) noted in describing these one-shot case studies, 'such studies have such a total absence of control as to be of almost no scientific value' (p. 6).

Why would scientists use this design, given its fallacies? Shadish et al. (2002) argued that when the 'effect is large enough to stand out clearly, and . . . the possible alternative causes [are] known and to be clearly implausible' (p. 107), then this design can have value. It is also possible to use a theoretical benchmark based on *causal* observations. For example, a broad expectation of what the data on the participants would have been had the stimulus not taken place – like the use of a parachute to prevent major trauma related to gravitational challenge (Smith & Pell, 2003). These designs can also be used by scientists conducting preliminary investigations of efficacy (Travis, 1983).[1] They may also serve as the basis for process evaluations and qualitative work (e.g. J.A. Greene, 1999; Irving & Hilgendorf, 1980), or to lay out hypotheses for future research. For example, to test whether exposure to televised sexual violence in crime dramas is associated with attitudes related to sexual violence, one may initiate the research project using a before-and-after design.

---

[1]As highlighted by Singal et al. (2014), efficacy is defined as the performance of an intervention under ideal and controlled experimental conditions, whereas effectiveness refers to its performance under 'real-world' experimental conditions.

Furthermore, as Lundivian et al. (1976) correctly noted that

> the numerous problems associated with this type of pre-experimental design make it extremely difficult to evaluate the results of the treatment effort. . . . It is impossible to firmly determine whether 'observed' or measured changes are a result of exposure to an independent variable or the result of 'uncontrolled', extraneous variables such as history, maturation, or regression. (p. 301)

Therefore, the one-group, post-only design remains a common design but should only be applied with a proviso that the study lacks the necessary precision to guide practice and inform us about the causal relationship (between the independent and dependent variables). Our position is therefore that its value both for scientific knowledge building and for policy implications remains limited.

The one-group pre-test–post-test design

Another design that is deemed 'bad' (Campbell & Stanley, 1963, p. 7), and was ruled out as the foundation for national recommendations on crime prevention (Sherman et al., 1998), is the before–after model. Such studies are likely to be applied most commonly worldwide to illustrate the 'success' of interventions on a group of exposed participants (Michel, 2017), especially in reports released by the research and evaluation departments of governmental offices (e.g. Israel Police Service, 2019). Most experimenters would harshly criticise this design, as it fails to adequately control for alternative hypotheses and, by implication, suffers from serious concerns regarding its internal validity (Farrington, 1983; Maltz et al., 1980; Marsden & Torgerson, 2012). Given their popularity across many disciplines (e.g. Bacchieri & Della Cioppa, 2007, pp. 183–199), we must pay close attention to the ways in which this design suffers from methodological limitations (see Chapter 3 regarding the threats to internal validity). The conclusion of this review ought to inspire reservations among readers from trying to test treatment effectiveness using this design. It should further serve to inform researchers about the danger in implementing policies based on this type of evidence. Results of these studies are useful for framing the research questions and hypotheses of future studies, but usually not as guiding evidence for practice.

Here, a researcher observes the units before they are exposed to the stimulus, and then measures the same units again after the stimulus has been administered. For example, in a study on the effect of a police training programme aimed at improving police–public relations, the police officers can be measured (i.e. tested) on their knowledge, skills, perception or behaviour prior to the module to form a baseline, and then measured again after the module has been completed (Israel et al., 2014). Any variation between the two time points (pre and post) is hypothesised to be a result of the training module. A gain in scores would be interpreted as a 'success', while a reduction from the baseline will be viewed as a 'backfire effect'.

However, a lot can happen between the two observation points in time, which may have caused the variation in scores. To illustrate why the pre-test–post-test designs are indeed 'poor', we will consider two threats that such experiments usually fail to control: history and maturation. Nonetheless, nearly all risks to internal validity are at elevated levels in this research design, including regression to the mean, testing and instrumentation.[2]

Historical events interfere with the causal inferences postulated by the experimenter and offer rival explanations for the differences, or the *delta*, between the two measures. When evaluating training programmes, for example, negative publicity involving police officers can have a meaningful effect on their confidence in their authority (Nix & Wolfe, 2017). If exposure to these high-profile cases occurs between O_1 and O_2 (i.e. observation at pre- and at post-stages), then we can no longer be sure that a change from O_1 is a result of the training module, or the negative publicity of the police. Clearly, the longer the temporal gap between the two measurements, the more likely those other, extraneous factors have affected the variation, but when considering the complexities of life, there are countless rival historical hypotheses that can debunk, exacerbate or reduce the treatment effect, to the point that we cannot rely on the trial results, even at short-term spans. It is not possible to achieve experimental isolation in field settings, and these confounding factors affect the conclusions without a reasonable method to control for them.

Similarly, there may also be *maturation* effects – the psychological and biological changes that participants may go through between O_1 and O_2 – which, again, confound the effect of the tested intervention. For example, Crolley et al. (1998) evaluated an outpatient behaviour therapy programme by examining 16 child sexual molestation offenders in Atlanta, Georgia. Using various psychological batteries and recidivism rates measured before and after completing treatment, the data were deemed to support the intervention. However, numerous systematic, naturally occurring internal changes – for example, spontaneous regression, natural mellowing of participants and so on – may have resulted in the before–after differences, rather than those differences being a result of the intervention.

---

[2]See a description of these threats to internal validity in Chapter 3.

We stress that these can take place at short-term durations as well. Unfavourable trainers who affect certain participants, poor settings (too hot or cold) and day-of-week or hour-of-day effects can all affect the personal engagement, attention or overall attitude of participants (e.g. Danziger et al., 2011), and they are all possible within-group variations outside the scope of the causal relationship tested in the trial.

Common tests of significance

In terms of analysing the results of a pre-test–post-test design, researchers have to consider several factors, including the distribution of the data (Are the scores spread normally? Were the participants recruited using a probability sampling technique? What is the level of measurement of the data?), sample size and whether there is access to data on individual participants (see De Winter, 2013). Under standard conditions, a *paired samples t-test* is useful, which calculates whether the mean difference score between two sets of observations (before and after) is statistically significant. In a paired samples *t*-test (sometimes also called a *dependent samples t-test*), each participant is measured twice, resulting in pairs of observations. The participants in the previously mentioned training programme, for example, would be measured once prior to and again after completing the training module, and the differences between the scores would then be analysed using this statistical procedure (which is widely available in all standard statistical software packages).

There are fundamental assumptions that accompany the paired samples *t*-test – such as the variable being continuous, the observations being independent of one another, the normal distribution of the data and the absence of systematic concerns with outliers (Weisburd & Britt, 2014, pp. 288–292). When these assumptions are not met (and mathematically accounted for using standard statistical procedures), then analogues to the *t*-tests are required. Similarly, if the researcher cannot match the pre and post scores for each participant, then the *independent samples t-test* will have to be used (Weisburd & Britt, 2014, pp. 270–287). It may be the case, for example, that the identity of the participants is undisclosed to the researcher, or purposefully anonymised by the treatment provider to avoid violations of privacy. Performing an independent samples *t*-test on scores that are in fact dependent on the baseline violates the test's assumptions (as well as reduces the statistical power of the test; see T.K. Kim, 2015, p. 544), and alternative, non-parametric statistical tests (Siegel, 1957) might be required.

The static-group comparison design

The static-group comparison design is a methodological framework in which the group exposed to the experimental stimulus is compared to another unexposed group. Both groups are only observed at the post-test stage. Put differently, the measurement of scores is taken after the intervention, without any baseline measures from the treated or untreated group, and without a mathematical attempt to create similar groups. For this reason, Shadish et al. (2002) refer to this as the post-test-only design using 'non-equivalent' groups: the reference to non-equivalence describes the unestablished differences between the groups in their baseline outcome measures.

For example, researchers may want to understand how police officers who are involved in community-oriented policing perceive their job performance (Sytsma & Piza, 2018). Such a study can take advantage of the geographic spread of the police and measure the intervention in a selected number of divisions, or even organisational units within the same division that engage in different policing styles. However, as many units can be considered unique in terms of their policing operations, it can be difficult to find proper comparison groups. The experimenter has no control over the assignment of the units either. Instead, scholars may choose to use a non-equivalent group comparison, at post-treatment, from within the same division – but without any pretreatment measures that can be used to establish the equality between the groups. Thus, this is an 'after-only' design, without randomisation.

Nonetheless, having no proper comparison group as would be the case in a 'true' experiment can mean that there is no 'formal means of certifying that the groups would have been equivalent had it not been for the [intervention]' (Campbell & Stanley, 1963, p. 12). The researcher does not have the ability to assume, or quantify, that the two study groups are equal in other measures outside of the treatment effect that is delivered to the intervention group. The issue then becomes a matter of an unresolvable selection bias, which indicates that the two observations – for example, the units with community-oriented policing initiatives and those that delivered a different style of policing – are not comparable at baseline. The researcher must therefore 'take a leap of faith to say that comparing post-test scores provides evidence of time order and an association, let alone controls for internal threats to validity' (Engel & Schutt, 2014, p. 132). More importantly, 'matching on background characteristics other than [the observation post-treatment] is usually ineffective and misleading, particularly in those instances in which the persons in the experiment group have sought out exposure to the [intervention]' (Campbell & Stanley, 1963, p. 12).

Common tests of significance

Again, researchers have to consider several factors (data distribution of scores, recruitment patterns, level of measurement, sample size, etc.). Under standard conditions, a *chi-square* or an *independent samples t-test* is useful. As there are no pretest measures, it is not possible to quantify the degree of sameness between the treated and the untreated participants, so no covariates of pretreatment conditions can be utilised.

True experimental designs

The pre-test–post-test control group design

This design is deemed the 'classic' experimental design, with two parallel groups measured at pre-test and then at post-test, following the random assignment of units into the different study groups. As this design is the most famous and is highly associated with the concept of an 'experiment' that soundly addresses the internal validity threats discussed in Chapter 3, we will spend more time on this design than any other. We will first review how the design approaches internal validity and then how external validity can remain a concern for experimenters who are considering the employ of this design.

In this design, a sample or a population of eligible units is first selected for the purpose of the study. This group of people, cases or other types of units of analysis are recruited into the study because they are characterised by a particular defining feature and the researcher hypothesises that exposure to a treatment will affect that feature. For example, a sample of offenders are recruited to a study that looks to reduce their involvement in crime, for example, through Cognitive Behavioural Therapy (CBT) (Barnes et al., 2017), restorative justice conferences (Sherman et al., 2015; Strang et al., 2013) or family violence interventions (Mills et al., 2019). In the language of experiments, these are the independent variables (i.e. the intervention), while the involvement in crime, compliance or victimisation would be the dependent variables. Case Study 4.1 lists some of the leading recent experiments that observed effects of these independent variables in the context of the criminal justice system.

(We should emphasise here the distinction between the random selection of units and the random assignment of units in the experiment. Random selection is related to the procedure of sampling – so its application is mostly in the context of external validity. The idea of random selection is derived from probability theory as well: a process of selecting a sample that is representative of the larger group from which it was drawn. Random assignment, on the other hand, is associated with internal validity. It is also derived from probability theory, but its aim is to help assure that the treatment and control groups are equivalent prior to the treatment.)

Once recruitment has concluded, consent for participation has been collected (whenever possible, see Vollmann & Winau, 1996; Weisburd, 2003) and the settings are optimal for assigning participants to the various study groups, the experimenter then conducts random assignment using one of the allocation protocols discussed in Chapter 2. Then, the experimenter would measure the dependent variable in each group at baseline – or O_1 for the treated group and O_3 for the untreated group see design 4 in Figure 4.1. The studied treatment is subsequently applied to the group that was assigned exposure to the treatment stimulus (CBT, retroactive justice conferences or therapy), but not to the other groups. Finally, another measure is taken of the dependent variable at the post-test observation stage (O_2 and O_4, respectively) of both the treated and the untreated participants.

Overcoming the risks to the internal validity of the test

Due to randomisation, we anticipate that the two groups are comparable at this pretest observation stage (O_1 and O_3). Having a no-treatment group that is similar to the treatment group at baseline confirms that many of the threats to internal validity are controlled for under this design. Similarly, since there is a follow-up of units from T_1 to T_2 in the no-treatment group, the researcher has a valid counter-factual estimation of what would have happened to the participants over time if the treatment were not administered. By comparing the dependent variable at the pre-test observation and then at the post-test, experimenters are able to tell whether the independent variable caused a variation in the dependent variable, *relative* to no-treatment conditions. Thus, the mechanisms for making a valid causal inference are satisfied.

However, in the social sciences, it is far more accurate to say

> the comparison of X with no X is an oversimplification. The comparison is actually with the specific activities of the control group which have filled the time period corresponding to that which the experimental group received the X . . . that these control group activities are often unspecified adds an undesirable ambiguity to the interpretation of the contribution of X. (Campbell & Stanley, 1963, p. 13)

Notwithstanding this ambiguity, the most appealing feature of the classic experimental design is that it neatly controls for rival hypotheses. The effects of *history* are controlled for, because any effect that may have occurred in the treatment group necessarily caused an effect on the participants in the control group. The variation from O_1 to O_2 in one group and then from O_3 to O_4 in the other group due to history is equal in the two groups, because both are systematically and simultaneously exposed to the same factors. For this reason, it should be immediately clear why a pre-test–post-test without a control group design is unable to control for history, whereas an experiment with two parallel groups can.

Note, however, that controlling for history also requires a regulation for simultaneity. As we suggested earlier, if the experimental group is run before the control group, or vice versa, history effects remain a concern: different times of the day or days of the week may cause the groups to be systematically different (Danziger et al., 2011). The optimal solution for this problem (amongst others discussed below) is to randomise experimental occasions – that is, therapy sessions, moments of exposure to the intervention or participation in individual training sessions. If there are extraneous history effects, they will be equally distributed across the experimental and control units (Leppink, 2019, pp. 248–249). All those in the same occasion share the same history, and therefore have sources of similarity other than the intervention.

In a similar way, *maturation*, *testing* and *instrumentation* effects are also controlled for using this design (Campbell & Stanley, 1963, p. 14; Shadish et al., 2002, pp. 257–278). Any psychological and biological factors are distributed equally in both groups due to randomisation. Importantly, both manifested and latent factors are similarly distributed in the two groups – both measurable and unmeasurable variables and those that were measured and that went unmeasured by the research team.[3]

Similarly, utilising the same fixed measurement instrument in the two groups is expected to result in comparable overall scores if the treatment condition were not applied. Had there not been a stimulus, the pre-test and post-test measures should produce similar results. One exception to this neatness is the use of a relatively small number of observers, in a way that makes it impossible to randomly assign them between the experimental sessions. For example, in policing studies, researchers often deploy observers in systematic social observations like ride-alongs with the police to document interactions between officers and citizens (e.g. see Berk & Sherman, 1985; Hirschel & Hutchison, 1992; Sherman & Berk, 1984, p. 264). However, these sessions can become very expensive (paying for assistants' time is a major cost) and complicated (Sahin, 2014), and usually only a small research team is available to conduct these observations. Therefore, to reduce any bias associated with one but not other observers, they ought to be randomly assigned to different

---

[3] To clarify, the 'measurable' variables refer to information that exists in the data available for the researcher to analyse, whereas the 'unmeasurable' variables refer to possible confounders that affect the results, but for which no data are available for the researcher to measure and to use in the analysis.

occasions, and to 'double-blind' them as to the randomisation sequence. This is not always as achievable in field settings as it may be in clinical trials, which can be a source of concern. At the very least, interrater reliability measures are needed to empirically estimate this source of potential bias,[4] though they are often not conducted (Jonathan-Zamir et al., 2015).

Proper randomisation also takes care of the risk of *regression to the mean*. Recall (Chapter 3) that the issue of regression is particularly concerning when the pre-test scores are extreme. After all, interventions are usually applied in these very settings: the most prolific offenders, hotspots with the most recorded crime, victims who suffered the most harm and so on. However, these units are the most likely to regress to the overall group mean over time. But with random allocation, possible regression effects are spread equally between the treatment and no-treatment groups – including outliers in the data (J.N. Miller, 1993).

Similarly, selection biases are greatly reduced in the classic experimental design, again owing to randomisation, which creates two parallel groups that are similar at baseline. However, we emphasise that pre-test inequality remains a possibility, as we discussed in Chapter 2, despite the random allocation of units. Selection effects persist as a concern even in large studies, but the problem is not ubiquitous in all experiments. In practice, this means that for individual trials there is a chance that statistically significant differences will emerge at the pretest scores, especially in terms of the dependent variable. The randomisation can allocate the participants into treatment and control conditions, 'assuring the unbiased assignment of experimental subjects to groups', but it does not guarantee 'initial equivalence of such groups. It is nonetheless the only way of doing so, and the essential way' (Campbell & Stanley, 1963, p. 15). Importantly, matching techniques are not able to overcome this issue, and in fact can be counter-effective: they lull the experimenter to think that they have created equivalence among the groups, when in fact this is not true. There are more valid procedures than ad hoc quasi-experimental matching techniques, as we detailed under blocking, minimisation and stratification processes (Lum & Yang, 2005; Nagin & Weisburd, 2013; Weisburd, 2010).

One issue that is not precisely dealt with using the classic experimental design is *mortality*, or the attrition of cases over time. In field experiments, scholars should expect some cases to drop out, especially among chaotic populations such as recidivist offenders, busy administrators or demanding probationers. By itself, this is not a source of concern – given the random allocation of these units into treated and

---

[4]One method for measuring how well raters agree with each other, and the consistency of the rating system, is the intercluster correlation coefficient. It is a widely used reliability index in test–retest, intrarater and interrater reliability analyses (Koo & Li, 2016).

untreated groups. Mortality becomes a source of concern, however, when it systematically characterises the treatment more than the control groups. For example, if the treatment requires the engagement and participation of the experimental units, but not the control units, then the latter is at a greater risk to drop out over time. This may be particularly the case in experiments that take many months to complete. For example, Sherman and Weisburd (1995) reported on a design breakdown towards the end of their influential hotspots policing experiment, with 'virtual disappearance of a difference in patrol dosage between experimental and control groups in the summer months' (p. 639). Thus, there is a link between the impact of implementation fidelity on mortality, and mortality can be a source of concern if it systematically affects one group more than others, in the pretest, post-test or treatment stages.

The solution to the unequal distribution of mortality is *not* the removal of the dropouts from the analysis, unless the same is repeated in the other group (see conditions in Fergusson et al., 2002). The problem, however, is that there may be no dropouts in the control group, when the study assumes true counterfactual with nil treatment in the comparison group. Finally, the risk of biasing the test fidelity in favour of the treatment group overshadows the benefits of potentially creating balance in terms of mortality effects. Consequently, any procedure in which non-completers are removed from the data can confuse the interpretability of the results, and therefore should be avoided. On the whole, we are in favour of the ITT[5] model, which albeit more conservative, provides the most stringent chance of avoiding these biases (Hollis & Campbell, 1999). It is ill advised to remove cases as an analytical strategy because the motivation for the mortality is usually indeterminable or difficult to quantify (Antrobus et al., 2013).

---

[5]As per Chapter 3, the ITT model is an analytical framework in which all units randomly assigned to the study groups are included in the final outcome analysis *as if* they completed the trial as assigned – even if they dropped out, died, switched study groups or were exposed to a treatment that was not initially assigned to them. ITT is particularly informative when the tested intervention is a certain policy *offered* to people, because in real-life settings, people do tend to drop out, die and switch between policies. Due to the random allocation of units into the experimental arms, the pretreatment conditions for not complying with the experimental protocol are distributed randomly as well. Therefore, ITT freely assumes that dropping out is part of the package, but with an equal probability of dropping out between the study groups. This balance allows the researcher to ignore these occurrences in the analyses and not treat it as a covariate or an endogenous factor. In practical terms, if 100 participants were randomly assigned into two even groups, then the denominator for each group, for any overall outcome of interest, would be all 50 originally assigned to the group.

However, there are conditions in which the dropping out rate can be extreme, and statistical corrections to the causal inference model should be considered (see e.g., Angrist, 2006). While these tools should be used very cautiously, if at all (Sherman, 2009, p. 12), they do present a viable solution for estimating and then correcting for dropping out effects.

Common tests of significance

Campbell and Stanley's (1963) view is that when the unit of analysis is a single person or case, then the two most acceptable statistical tests are to compute gain scores or to use tests such as analysis of covariance (ANCOVA). In the gain scores approach, scholars first compute a pre-test–post-test gain score for each unit (e.g. T_2 minus T_1), and then compute a t-test value between experimental and control groups on these gain scores. However, ANCOVAs with pre-test scores as the covariance are 'usually preferable to simple gain-score comparisons' (p. 23). When the unit of analysis is an intact group or cluster, then the group means (rather than the individuals' scores) should be used instead (e.g. Langley et al., 2020).

We note, however, that these common tests need to be applied when there is a logical assumption of representativeness, otherwise the fundamental purpose of conducting parametric statistical tests of significance such as the t-test or ANCOVA will not be met (Weisburd & Britt, 2014). In practical terms, unless the researcher has conducted probability sampling, or utilised the entire population of eligible cases, the assumption of representativeness will not be met. Using volunteers, cohorts of fixed populations or paid recruits has intrinsic value when the purpose is to generalise onto these very confined universes, to these populations only.

The post-test-only control group design

Interestingly, while experiments are famous for having pre-test measures, in principle they are unnecessary. We are accustomed to observing baseline equivalence and reporting on balance between the treated and untreated groups, but randomisation can suffice, without any pre-test scores, to account for inter-group differences other than the allocated treatment. Put differently, the pre-test measure may not be worth the trouble, and unless there are reasons to distrust the randomisation, a post-test-only control can be enough. For example, Ariel et al. (2015) tested the effect of body-worn cameras in policing, using the police shift as the unit of analysis. However, as the shifts do not represent a continuous population of units, baseline measures could not be taken. Thus, only a post-test measure was possible under these conditions. Differences between O_2 and O_4 are still assumed to result from the intervention due to the random allocation, but there is no way to quantify baseline equality in terms of the dependent variable of interest.

Moreover, there are real-life considerations that limit the ability of the researcher to collect pre-test scores. First, the researcher may be unable to measure the groups at pre-treatment stage (i.e. the participants are unavailable, their anonymity does not allow for a pre-test measure, there are no measurable data, etc.), or a pre-test measure can only be taken from the treated but not the untreated group because observing the no-treatment participants would undermine their no-treatment status. For example, control participants may react to the pre-test observation, and this reaction will cause their pretest measures to reflect awareness to the test in places where it is not wanted: a new policy, intervention or status associated with the intervention (Ariel, Sutherland, & Bland, 2019).

Common tests of significance

The simplest approaches would be to use an *independent samples t-test*; however, covariance analysis or blocking on participants' characteristics are preferred, as they increase the statistical power of the test. For an explanation of blocking, refer to Chapter 2.

The Solomon four-group design

In many ways, this is the pinnacle of a true experimental design, because it deals directly not just with internal validity threats but also with the issue of external validity linked to testing effects. As noted earlier, people are affected by the observation, not just the intervention – for example, they can be sensitised in the pre-test to being observed or tested, they may gain knowledge about the intervention from the pretest measure or they can alter their natural response to the intervention, due to the measure taken at baseline. For example, any test of aptitude, knowledge or skills is likely to suffer from a pre-testing effect, because participants have likely learned from the pre-test (Ariel, Sutherland, & Bland, 2019). To control for these, the Solomon four-group design (Solomon, 1949) was developed – however, we do note that it is rare in the social sciences (see exceptions like Fonow et al., 1992; and the DARE [Drug Abuse Resistance Education] evaluation by Dukes et al., 1995).

As Design 6 in Figure 4.1 illustrates, the Solomon design incorporates all four groups to which participants are randomly assigned: two parallel groups like the classic experimental design (both measured at pretest) and also two parallel groups that are measured within a post-test-only groups design (i.e. no pre-test, and only one of these two groups is exposed to the intervention). Once the sample has been fully recruited (or the entire population of eligible cases), all units are then randomly assigned into the four groups. Due to the randomisation, pre-test equality is assumed across all groups. However, there are situations when participants' perceptions are susceptible to the effect in the message, whether manifested or disguised in the pre-test measure. When we anticipate that pre-test measure O_1

may interact with the treatment X and the post-test observation O_2, and that the pre-test measure O_3 may interact with the post-test observation O_4, we can control for these interactions by establishing how the participants behave without these pre-test measures (i.e. O_5 and O_6 only).

Using this incredibly useful but underutilised model, the threats to internal validity are controlled, with particular emphasis on testing effects – but it also controls neatly for external validity threats. If across all comparisons the treatment effect remains consistent and pronounced, then the strength of the inference is increased.

Common tests of significance

t-Tests and the variations of the two-sample tests are not appropriate, as there are more than two groups. One recent innovative statistical test is to apply a meta-analytic approach (one which combines results from a number of independent studies to describe overall trends in effect), given the random allocation into the groups. For details, see Braver and Braver (1988).

Case Study 4.1

RCTs in the Criminal Justice System

RCTs have led eminent criminologists to convincingly show how various interventions cause outcomes in the criminal justice system context, relative to control conditions. For example, based on these experiments, it has been shown that

- directed police presence in crime hotspots reduces crime and disorder (Ariel, Sherman, et al., 2020; Ariel, Weinborn, et al., 2016; Ratcliffe et al., 2011; Ratcliffe et al., 2020; Sherman & Weisburd, 1995);
- face-to-face restorative justice led by police officers reduces recidivism, increases victims' satisfaction, lowers their post-traumatic stress symptoms and saves revenues, compared to usual criminal justice system processes (Angel et al., 2014; Sherman et al., 2015; Strang et al., 2013; see also Mills et al., 2013; Mills et al., 2019);
- innocent suspects are less likely to be mistakenly identified and guilty suspects are more likely to be correctly identified in simultaneous rather than sequential police line-ups (Amendola & Wixted, 2015);
- nudges in the criminal justice system do not usually lead to desired effects as the theory predicts, with non-significant differences between various reminders and control

(Continued)

conditions (Chivers & Barnes, 2018; Cumberbatch & Barnes, 2018; Monnington-Taylor et al., 2019), however see the Science paper suggests otherwise https://www.ideas42.org/wp-content/uploads/2020/10/Behavioral-nudges-reduce-failure-to-appear-for-court_Science.full_.pdf

- placing a marked police patrol in certain areas reduces property crimes (Ratcliffe et al., 2020);
- police body-worn cameras can lead to reductions in complaints against the police (Ariel, Sutherland, et al., 2017; Ariel, Sutherland, et al., 2016a, 2016b) as well as assaults against security guards (Ariel, Newton, et al., 2019) and increase the perceived legitimacy of the police (Ariel, Mitchell, et al., 2020; Mitchell et al., 2018; however, cf. Lum et al., 2020);
- police-led target-hardening crime prevention strategy to burglary victims and their close neighbours does not lead to statistically significant reductions in repeat or near-repeat burglary (Johnson et al., 2017);
- policing interventions directed at increasing collective actions with citizens at crime hotspots increase citizens' fear of crime (Weisburd et al., 2020);
- problem-oriented policing strategies reduce the incidence of violence in hotspots (Braga et al., 1999; Telep et al., 2014);
- procedural justice practices affect citizens' perceptions of police legitimacy, trust in the police and social identity (Mazerolle, Bennett et al., 2013; Murphy et al., 2014);
- requirement to attend brief group therapy with cautioning leads to reduced subsequent reoffending in low-level intimate partner violence (Strang et al., 2017);
- 'scared straight' programmes backfire (Petrosino et al., 2000);
- second response programmes (i.e. police interventions that follow the initial police call for service) to tackle domestic violence do not 'work' (Davis et al., 2010);
- standard or reduced frequency of mandatory community supervision for low-risk offenders does not lead to different recidivism rates (Barnes et al., 2010);
- training police recruits on procedural justice results in short-term benefits for police–public relations (Antrobus et al., 2019);
- truancy interventions in schools lead to reductions in violent behaviour (Bennett et al., 2018; Cardwell et al., 2019; Mazerolle et al., 2019);
- using civil remedies (i.e. non-offending third parties such as property owners) controls drug use and sale (Mazerolle et al., 2000) and
- working 10-hour shifts is healthier than working 8-hour shifts among police officers (Amendola et al., 2011).

Quasi-experimental designs

Why you might not be able to conduct a pre-experiment or true experiment

Prospective experiments are not always possible. There are several reasons why retrospective studies are needed to estimate cause-and-effect relationships. First, there are certain questions that cannot be dealt with prospectively because the exposure to the treatment, the outcomes and the allocation of cases have already occurred. Second, it may not be ethical to conduct true experiments (see Mitchell & Lewis, 2017; we discuss these issues in Chapter 5). Third – and perhaps most crucially – treatment providers are not always open to the idea of random assignment. From anecdotal experience, we can say that prison authorities, lawyers and police departments are often apprehensive about RCTs. Therefore, despite the limitations we have discussed, 'natural' experiments and *quasi-experimental designs* often fit the bill.

For example, testing the effect of allocating court cases to judges of a different ethnicity, with a view to measuring how different judges treat defendants of different backgrounds, is likely to be best studied as a **'natural experiment'**, as the allocation of cases has already happened (e.g. Gazal-Ayal & Sulitzeanu-Kenan, 2010). Such studies are still considered experiments because the distribution of court cases to the judges is done with a certain degree of randomness, that is, without a systematic pattern that prefers certain judges to others in the allocation of new cases – unless the case requires a judge with expertise in a subject matter. Under these conditions, we would be able to falsify the **null hypothesis** of no differences between the groups of judges and understand whether their backgrounds matter in terms of court outcomes.[6]

Far more common, however, are the retrospective causal studies within the 'quasi-experimental designs' category. In these studies, the researcher obtains *existing* datasets – police records, survey responses or court cases – about a particular phenomenon and is interested in observing a set of independent variables and their relationship with a set of dependent variables. The researcher is not involved in the allocation of the treatment, but still attempts to extrapolate the cause-and-effect relationship based on the exposure to a certain intervention in one subgroup of cases (offenders, officers, places, etc.) and then compare it to another subgroup of similar cases. The 'trick' in these designs is the creation of a comparison group that is equal to the 'treated' group, so that the claim of causal relationship is sufficiently compelling.

These *quasi-experimental* designs come in many forms, including regression discontinuity, interrupted time series and propensity-based methods (Angrist & Pischke, 2014; Cook & Campbell, 1979; Morgan & Winship, 2007, 2012; Shadish et al., 2002; cf. Shadish, 2013). These approaches tend to blur the distinction between 'design' and 'analysis', because they are inherently statistical in nature and are still

---

[6]Recently, some scholars have applied a 'synthetic group' experiment (Abadie & Gardeazabal, 2003) in situations where there is no comparison group running simultaneously with the treatment group. A recent and interesting example can be found in Bartos et al. (2020).

considered 'experimental' designs (Rubin, 2008). Below, we explain what each of these designs means, but more broadly, as recently reviewed by Ariel (2018, pp. 73–75), by employing statistical procedures of varying degrees of sophistication, quasi-experimental studies seek to exploit subgroups within the data that can serve as counterfactual conditions to the subgroup of cases that were exposed to the intervention of interest. At this juncture, however, many studies that use quasi-experimental designs are criticised: does the comparison group indeed represent a 'like-with-like' comparison?

If we consider the approach taken by Campbell and Stanley (1963), this is rarely the case. Modelling the world by secluding one factor that differentiates treatment from control conditions is just not possible in non-RCT settings (as we reviewed in Chapter 2). This means that scholars who are measuring the impact of policies ought to be aware that the treatment will always be confounded by the thousands of variables, not to mention the countless interaction effects that can take place in the background – unless randomisation is included (Ariel, 2018, p. 68). We can say that non-RCT models are asymptotically close to creating counterfactual conditions; however, only randomisation (with sufficiently large samples, across multiple trials) can create the conditions required to draw true conclusions about causality.

Still, these alternative procedures have real-world value (Shadish et al., 2002) and are frequently implemented, so they merit consideration. Here, we briefly set out the processes and problems associated with the most common alternatives that researchers turn to for causation exploration in the absence of randomisation.

The (interrupted) time-series experiment

The time-series experiment can be thought of as a more elaborate before–after design with one group, with multiple observations prior to the exposure to the stimulus and then multiple post-test observations after the exposure (rather than only two moments in time). This model should not be confused with the observational time-series design, which does not aim to deduce causality, but rather to understand the developmental progression of participants as a function of time. For example, scholars have followed public perceptions in the context of policing and crime to illustrate how confidence has changed over time (e.g. Sindall et al., 2012). In the context of causal studies, however, the **time-series analysis** is one of the most logically convincing models we have, especially in physics, biology and chemistry. For example, we can measure the room temperature every hour during a 24-hour period, and in T_{12} the thermostat is then turned up by 10 degrees (which would result in an increase in the temperature by 10 degrees and in every $T_{13...24}$). We can attribute the increase in the temperature in the room to the change in the thermostat, and nothing else; it is difficult (if not scary) to think of another factor why the temperature had risen in observations following T_{12}, other than the adjustment of the thermostat. This is particularly the case here, because we have a perfect dose–response relationship (a change in the thermostat of 10 degrees has led to a change of exactly 10 degrees in the room temperature, in a consistent manner).

This design suffices for straightforward physical expressions of cause and effect. The question, however, is whether the time-series experimental model is sufficiently strong to control for threats to the internal validity of quasi-experiments in the social sciences – and the answer is plainly no. We discuss these issues in Chapter 2 (and more robustly in Nagin et al., 2009; Nagin et al., 2015, p. 92; Sherman, 2009, p. 13), but we provide here an example to emphasise this issue. An experiment by Loftin et al. (1991) looked at the relationship between restrictions on firearms sales in Washington, D.C., and levels of violent crime. The study has shown that, following restrictions, a reduction in gun-related crime was recorded.

Though this cause-and-effect relationship is logical, the evidence of the drop in gun-related injuries following the change in rules limiting the sale and licensing of firearms can still be reasonably explained through rival hypotheses. Have gun-related injuries dropped in jurisdictions without gun restrictions? Are there scenarios where gun licensing restrictions resulted in no impact or even an increase in gun-related injuries? Has the increase in police efficiency in combating gang-related crime affected the frequency of gun-related injuries, but not the licensing rules? Finding comparable control conditions in these settings is very difficult, as the units of analysis are entire jurisdictions that can cover many millions of people (Toh & Hernán, 2008). Unlike the room temperature example, or any other experiment in nature where a variation in the data is incontestably a result of the stimulus, the social sciences do not 'work' this way. This *one*-group time-series analysis cannot control for alternative explanations to the apparent reduction in gun crime when gun laws are put in place.

However, there is one possible scenario in which an extension of this time-series experimental design would be considered sufficiently powerful to remove rival explanations: *multiplicities*. As we stressed, any *one* study utilising a time-series model is deemed insufficient, but a *series* of studies using time-series analyses, in different settings, with diverse populations and at different times, which collectively result in similar trends across the time-series, can produce convincing causal estimates. Indeed, these will remain asymptotic: it is a method that approaches the necessary conditions for true cause and effect, but never fully arrives at these necessary conditions. But multiple time-series experiments, from different sources, showing the same direction and magnitude of effects, can be quite convincing.

We note that, overall, this and similar studies were subsequently incorporated into a systematic review by Crandall et al., 2016, and Lee et al., 2017, on the link between restrictive licensing laws and firearm-related injuries, and the conclusions were similar: stronger firearm laws are associated with reductions in firearm homicide rates.

The multiple time-series design

To capitalise on the concept of multiplicities, an experimenter may use an extension of the interrupted time-series experiment by adding a second (or more) simultaneous time-series from a comparable control group(s). This design can provide some safety from the threats to validity of the single time-series experiment discussed above – especially history and maturation effects. Often referred to as a *controlled interrupted time-series design*, this model uses several waves of data in 'treated' and 'untreated' groups, pre and post the exposure to the independent variable in the treated group but not in the untreated groups (see Bernal et al., 2017).

In Chapter 3, we introduced Campbell's (1968) Connecticut crackdown on speeding time-series experiment, which is a classic example of this design. As we noted, multiple threats to internal validity may jeopardise this design – even though it provides more controls than the single-group time-series experiment discussed above. Still, from the various quasi-experimental designs that can be used to evaluate interventions, this is one of the most accurate, because it incorporates multiple measures as well as a comparison group.

A related (but not identical) model is the **difference-in-differences** design (DID). In a controlled interrupted time-series design, programme impacts

> are evaluated by looking at whether the treatment group deviates from its baseline trend by a greater amount than the comparison group. However, the DID design . . . evaluates the impact of a program by looking at whether the treatment group deviates from its baseline mean by a greater amount than the comparison group. (Somers et al., 2013, p. iii)

DID calculation shows causal effect as a function of the difference between the changes in the before and after means of the two groups (see Figure 4.2).

The design may prove useful in natural experiment circumstances (e.g. see Bilach et al., 2020). However, the major drawback of DID is that it requires a substantial set of assumptions about the comparability of the comparison and treatment groups at pre-intervention stages (known as the parallel trends assumption). It is unlikely that the research team can be confident that this design accounts appropriately for *all* confounding variables required for a causal inference, which therefore places limits on our confidence in conclusions from DID designs.

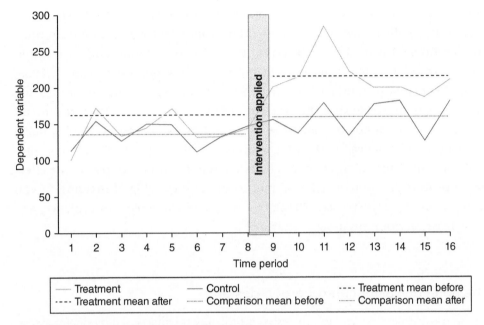

Figure 4.2 Typical difference-in-difference design

The equivalent time-samples design[7]

The equivalent time-samples design is also a one-group design, very similar to a single-study interrupted time-series experiment, but with multiple observations *and* multiple exposures to the treatment. In this design, the researcher introduces the treatment multiple times, and withdraws it at others. Observations of the outcome can then be taken after each exposure to the treatment and after each time where the intervention is withdrawn.

Villaveces et al. (2000) applied this design to evaluate the result of intermittent police-enforced bans against carrying firearms on the incidence of homicide in the cities of Cali and Bogota, Colombia. Carrying firearms was banned on weekends after paydays, on holidays and on election days. Enforcement included 'establishment of police checkpoints and searching of individuals during traffic stops and other routine law enforcement activity' (p. 1205). This quasi-experimental design concluded that an intermittent ban was associated with a reduction in homicide rates for both cities.

One issue with this design is its external validity: rarely are people repeatedly exposed to multiple levels of the intervention. We are only able to generalise

---

[7]A related model is the *equivalent materials design*, which includes different treatment materials; its rationale and design are ostensibly the same as the one presented here.

findings from such an experiment to other settings in which the intervention is organically applied repetitiously. It would be a mistake to generalise from these rare conditions – switching the intervention on and off intermittently – to conditions in which this intervention only appears once (Campbell & Stanley, 1963, p. 44). There is also an interaction effect between the experimental occasions and the observations, or between the different observations – in ways that are difficult to quantify (you cannot separate these autocorrelations). Therefore, the Colombia evaluations are practically useful for settings in which a gun ban is not continuously present. The findings ought to be assessed in the wider context of the effect of police strategies to reduce illegal possessions and carrying of firearms on gun crime (Koper & Mayo-Wilson, 2012), but as a stand-alone piece of evidence, any generalisation from this design is narrow.

The non-equivalent control group design

The non-equivalent control group design is probably the most-used model for assessing causality in quasi-experimental settings – and programme evaluation research usually takes this approach. The design looks identical to the pre-test–post-test control group design, but without the random assignment of units into groups. In other words, both the treated and the untreated groups are observed once at pretest and again at post-test; however, there is no pre-experimental equivalence through randomisation. The groups are deemed comparable by virtue of belonging to the same overall population. The greater the similarity between the experimental and the control groups *confirmed* at the pre-test observation, the more effective the control then becomes.

Various statistical methods exist to measure the relative gain scores or the ANCOVAs (we favour the latter for efficiency purposes, as we discussed earlier when considering analytical approaches for true experimental designs). Through these procedures, the experimenter attempts to create a pre-test balance between the groups so that the only meaningful difference is the isolated intervention effect in the treatment but not the control group, using statistical controls. However, as we conveyed earlier, matching procedures with non-equivalent groups are categorically asymptotic to creating balance by using randomisation: they ought to be used when randomisation is not possible, but not as the first option (Campbell & Boruch, 1975).

One issue that strikes us as underexplored is the regression to the mean effects on both the measured and the unmeasured variables. To consider the severity of this concern, we must remember that many, if not most, experiments are performed on participants with extreme scores: the hottest hotspots, the most harmful felons, the most frequent reoffenders, the most harmed victims or those otherwise most in need of an intervention (Dudfield et al., 2017; Liggins et al., 2019; Sherman et al., 2016; J. Sutherland & Mueller-Johnson, 2019). Based on these extreme scores, the control group is then selected for matching purposes; after all, the selection of the comparison group must at least be done in a way that matches with the treatment group based on the dependent variable, to create the pre-experimental equivalence of groups. However, as we have shown in Chapter 3, there is a natural tendency for patterns to regress to the mean. If the means of the two groups are substantially different and require a matching procedure, then the process of matching 'not only fails to provide the intended equation but in addition insures the occurrence of unwanted regression effects' (Campbell & Stanley, 1963, p. 49). This implies that the two groups will differ on their post-test scores independently of any effects of the treatment.

Another issue is the self-selection of treatment group participants, as they either deliberately sought out or were 'forced into' exposure to the treatment, whereas the control group participants are deliberately not. This makes all the difference in the world: assuming that the participants share the same motivations, perceptions, prognosis or recruitment opportunities. For this reason alone, experimenters should be concerned about relying on findings from this design – unless these issues can be factored out. Often, this is not the case (but cf. Hasisi et al., 2016; Haviv et al., 2019; Kovalsky et al., 2020; G. Perry et al., 2017).

The separate-sample pre-test–post-test design

It is often the case that all units must undergo treatment at the same time, for ethical or organisational considerations. Under these conditions, it might still be possible to conduct an experiment to measure the treatment effect. Here, the participants are *randomly* assigned into treatment and control conditions to create pre-experimental equality. However, one sample is measured prior to the intervention and the other group is measured after the intervention, but no group is measured twice. This *separate-sample pretest–post-test* design is superior to the simple before-and-after experiment, as it has a control group; however, it is a lesser design compared to a true experiment as many of the threats to internal validity remain uncontrolled, including history effects (events can happen during the temporal gap between the pretest and the post-test observations), instrumentation effects (using the same observers in the pretest and in the post-test can cause variations in methods and expectations – although this may largely depend on the length of the temporal gap) and mortality (if the gap in time is wide).

Case Study 4.2

ThinkUKnow: Online Safety Training in Australia

A recent experiment addressed the effect of a police-delivered intervention in schools to educate students in online safety (Alderman, 2020). This 'ThinkUKnow' programme was provided to classroom groups in clusters of primary and secondary classes in Australia. The half-day training module was delivered by police officers, and the measurement included a questionnaire that assessed protective behaviours, knowledge, willingness to report victimisation to the police and perceived severity of online crime. The study reported significant score gains in the treatment compared to the control group.

From a methodological perspective, this experiment follows the separate-sample pre-test–post-test design (although it is somewhat of a 'patched-up design' as it also incorporates the blocking and clustering methodologies discussed in Chapter 3). Classes were randomly assigned into treatment and control conditions, where all pupils were exposed to the training, as it was a requirement of the education system. The availability of police officers to deliver as well as to administer the pre-test–post-test measurements was limited, so officers had to attend each school/class only once. For these two reasons, the control group (i.e. classes) was measured prior to the exposure to the educational model (which everybody was exposed to at the same time), and the treatment group was measured following the training – that is, there were no simultaneous pretest and post-test measures, but split, as this experimental design dictates.

The issue, however, is that this design forces the researcher to make assumptions about group comparability, and not all of these assumptions are sufficiently convincing. History effects are the most concerning. We assume that processes that the treatment classes have gone through would be fairly similar to the processes that the control classes have experienced, except for the stimulus. However, as we alluded to earlier, seasonal effects can be short term as well (participants' attention depletion can vary in the morning, as may the instructors'). This design also assumes that there is no testing crossover effect: control participants who were observed prior to the educational module sharing with the treatment participants did not share with those who were measured in the post-test only. This, again, is not necessarily the case, especially with students who have mobile phones and access to social media. Finally, and as importantly, mortality is likely to differ as well: the number of completers of the instrument is dependent on time, with more participants taking the pretest rather than the post-test assessment.

Regression discontinuity design

A regression discontinuity design (RDD) allows researchers to retain control over the assignment to the treatment condition and draw meaning from a comparison with a separate group that has not been exposed to the treatment. This is achieved by calculating a regression model for each treatment condition, where the conditions are completely separated by a continuous assignment (running) variable with a given threshold (for further technical details, see Berk et al., 2010, as well as Heinsman & Shadish, 1996).

The process is best demonstrated with a hypothetical example. Suppose a consortium of public agencies wishes to assess the effect on repeat offending by assigning high-risk domestic abuse cases to a multi-agency unit. We can capitalise on the fact that there are different degrees of 'high-risk' levels posed by the offenders, and the variable that defines the varying levels of risk posed by the individuals can be construed as an 'assignment' variable. The threshold variable is a proxy for an independent variable, as it stands in for the assignment of a case to the intervention. All those cases scoring above the threshold can be deemed the treatment group, and all those under the threshold are the comparison group.

RDD is typically presented graphically, in the form of several plots, the main of which is a scatter plot demonstrating the case relationships between the dependent variable and the assignment variable. The main function of the RDD procedure is then to apply a regression model to estimate the best fit. The resulting regression is then displayed with a smoothed line on the scatter plot, as shown in Figure 4.3. The threshold, indicated by the vertical dashed black line, demarcates the treatment group from the comparison group. The difference between the fitted values of the regression lines at this threshold point is the measure of the differential effect (Y. Kim & Steiner, 2016).

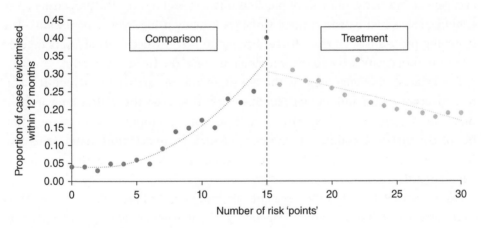

Figure 4.3   Example of regression discontinuity design

However, RDD is a proxy in lieu of randomisation only when two conditions are met: first, that there is continuity around the threshold point, and second, that all participants comply with their 'assigned' treatment condition. In practice, these conditions may not always be observed, and they interfere with the internal validity of RDD. Finally, RDD tends to have issues with external validity. Since RDD only estimates the treatment effect at the threshold, any generalisation for cases further away from this threshold level is compromised (Y. Kim & Steiner, 2016).

Propensity score matching

When RCTs are not feasible, matching is required, but the method of matching is not straightforward because researchers must correct for baseline imbalances. When access to covariates and the assignment process is provided, then researchers can use the procedure known as **propensity score matching** (PSM). PSM creates a custom control group by matching each unit of the treatment group to a control unit, which has the same or similar underlying covariates (for a detailed review, see Apel & Sweeten, 2010; Haviland et al., 2007).

Using PSM, scholars can quasi-experimentally test the effects of various crime-control interventions – from prison-based treatments, through hotspots policing, to community-oriented treatments (Grommon et al., 2017; Haberman et al., 2018). Though not without limitations (Rosenbaum & Rubin, 1983), PSM is particularly effective when researchers have limited cooperation from treatment providers to randomly allocate cases into treatment and control conditions (Hasisi et al., 2016; Weisburd et al., 2017).

Consider an example in which non-violent offenders are assigned to a CBT intervention. If researchers evaluating this initiative had access to data on the baseline characteristics of the individuals who were assigned to CBT and those who were not – that is, age, gender, ethnicity, number of previous offences and so on – the propensity score could be calculated based on these variables (usually using logistic regression). The matching process is such that the control group is constructed of the units that are closest in propensity score to the individual units in the treatment group.

The greatest challenge here is the ability of the researcher to include the correct covariates. Kim and Steiner (2016) suggest that deep theoretical knowledge is highly important to this aspect and caution against simply increasing the number of covariates to counter the absence of such theory. Unfortunately, the confounding assumptions of PSM are rarely explored in depth in practice (Shadish & Sullivan, 2012).

PSM, as with RDD, offers the researcher alternative techniques to explore causation in the absence of randomisation. Depending on a researcher's ability to identify confounding variables, these procedures are potentially effective – particularly in cases where it is impracticable to run a randomised experiment. However, it also must not be ignored that the design cannot offer a strong statement about cause and effect in the same way as randomisation. All experimental statistical techniques that do not utilise randomisation assume that all confounding variables have been accounted for, and as this chapter has shown, certainty in this regard is a rarity – any variables that have not been measured therefore cannot be incorporated in the statistical model. Attention to these concerns was reviewed in our discussion on the benefits of randomisation, in Chapter 2.

Conclusion

Not all experimental designs are created equal, and they vary in terms of the degree of control they exercise over the experimental process. We reviewed the 11 classic types of experimental designs laid out by Campbell and Stanley (1963) and two specific quasi-experimental innovations (RDD and PSM). We have made the case for implementing true experiments over other designs. When presented with the choice of either conducting an RCT or implementing a different causal design, RCTs should be preferred, for all the reasons we discussed in this chapter.

Our preference towards RCTs is not just about the benefits of randomisation for the creation of optimal counterfactual conditions. As we reviewed in Chapter 2 and then in more detail in Chapter 3, science has yet to develop a more convincing causal inference model other than the RCT. Indeed, experiments using random allocation are not always feasible when conducting field tests. They can be difficult to handle: case attrition, diffusion of treatments, inconsistent treatment fidelity and problems associated with small samples are common features. The experimenter ought to have a strong protocol to follow, as we discuss in Chapter 5, and conduct a robust pre-mortem analysis to plan ahead and then offset issues that will arise during the experiment. Still, true experiments have in them a set of conditions that protect the validity of the trial from a long list of threats, as listed in Chapter 3. The other experimental designs we reviewed in this chapter have more limited control. Therefore, when the conditions are possible, RCTs should be implemented.

Of course, this is not to say that other experimental designs have no value. As the body of evidence accumulates on any particular intervention, the pre-experimental designs can become informative as well. Why should we discount pre–post-only studies if a series of them show the same result, across different populations, settings and time? When these designs present outcomes that do not conflict with or strongly deviate from the overall body of evidence, they should not be ignored. Our knowledge of the tested intervention will only get richer, not more confusing, with the accumulation of evidence gathered from a range of research designs.

Finally, quasi-experimental designs, or statistical modelling on retrospective data, will always be required. Even though they have inherent issues associated with weaker control conditions to which the tested intervention is compared (e.g. selection bias), these models can provide great insight on tested hypotheses. Data sets created by the state (e.g. crime data), historic data and secondary data analysis more broadly will continue to provide opportunities to sharpen our theories of causal inference, using quasi-experimental designs. For these and other reasons, this chapter does not ignore non-RCTs at all: it celebrates their contributions, and instead focuses on the settings in which they are the most optimal.

Chapter Summary

- In this chapter, we examine 13 different types of experimental designs. As other classic textbooks have done before us, we catalogue these experimental designs into three broad categories based on each design's ability to exert control over the experimental process.
- First, there are true experimental designs that incorporate the random allocation of units into treatment and control arms of the trial (RCTs). These are considered the gold standard for causal estimations when all things are equal.
- Second, there are pre-experimental designs, which have some but not all features of the RCT. These include, among others, the before–after-only design, which are known as 'lesser' experiments because of the inability of the researcher to control for many of the threats to the internal validity of the test. They are nevertheless deemed 'experimental'.
- Third, we have quasi-experimental designs, which are usually used as retrospective experiments that create counterfactual conditions using statistical modelling techniques rather than randomisation procedures. Our descriptions favour true experiments over those with weaker validity, but any experimentalist must face the reality that a classic 'true experiment' is not always possible.
- As such, we offer a summary of alternative methods, highlighting their relative strengths and weaknesses in terms of threats to internal and external validity. A taxonomy of experiments in the social sciences, with a focus on criminology, is provided as well.

Further Reading

Campbell, D. T., & Stanley, J. C. (2015). *Experimental and quasi-experimental designs for research*. Ravenio Books.

Many (if not all) textbooks on research methods in the social sciences that discuss causal research have cited Campbell and Stanley's (1963) manual on experimental and quasi-experimental designs. This survey of the topic provides the foundation for this chapter as well. For further details on the experimental methods presented here, as well as other niche or bespoke designs, the original survey of experiments (reprinted) should be consulted.

Weisburd, D., Hasisi, B., Shoham, E., Aviv, G., & Haviv, N. (2017). Reinforcing the impacts of work release on prisoner recidivism: The importance of integrative interventions. *Journal of Experimental Criminology*, 13(2), 241–264.

Apel, R. J., & Sweeten, G. (2010). Propensity score matching in criminology and
criminal justice. In A. Piquero & D. Weisburd (Eds.), *Handbook of quantitative
criminology* (pp. 543–562). Springer.

The discussion in this chapter briefly introduced PSM as one of the most popular
techniques today for creating statistically balanced arms in quasi-experimental
designs. Seemingly, experimenters evaluating interventions in prison settings have
enjoyed more experience with this technique than other areas in criminology. Some
examples are particularly noteworthy, such as this recent study by Weisburd et al.
on the benefits of work release on prisoner recidivism. For further reading on PSM
in criminology, this chapter by Apel and Sweeten should be consulted.

5

RUNNING EXPERIMENTAL DESIGNS

Chapter Overview

Recipes for experiments

One of the most important stages of scientific inquiry is the meticulous planning ahead of all the necessary steps, hazards and possibilities that may occur during the experiment. Unlike random discoveries, which have always been part of the scientific journey, running a strategic experiment presents various obstacles. Trials are meant to be methodical, disciplined and structured. In this sense, prospective studies are different from retrospective analyses of data because the prospective experimenter cannot go back in time and reformulate the research hypotheses, or switch to a different intervention midcourse. Field and clinical trials usually have to 'stick' to the original game plan (see Sherman, 2010). Deviations often necessitate a restarting of the trial. As experiments tend to be expensive, and changes are often difficult to explain without jeopardising the integrity of the test, the recommended course of action is to act according to the plan. If nothing else, an experimental cycle that does not follow its own guidelines is a messy test to analyse and should therefore be avoided.

As alluded to in the synopsis, there are three issues at stake here. First, experimenters should plan how they will conduct their experiments using a protocol or template. Second, there are organisational frameworks that are more conducive to the efficient implementation of the experimental process. Finally, there are crucial ethical considerations to consider. We discuss these issues in this chapter and begin by exploring experimental protocols.

Experimental protocols and quality control tools

Researchers can use protocols, or blueprints, to help them conduct sound and robust experiments. In many ways, the protocol – a detailed plan of the overall experimental process – nudges the experimenter into compliance with the study design and, by implication, industry's guided practices. Experimental protocols are 'key when planning, performing and publishing research in many disciplines, especially in relation to the reporting of materials and methods' (Giraldo et al., 2018, p. 1). Protocols are important, especially in the form of checklists, because they serve as active aids and memoirs (see review of nudges and checklists by Langley et al., 2020). The closer the experiment sticks to the original recipe and follows its game plan, the more credible and transparent the results. This is true for the early planning stages, as well as the stage at which the final reports are disseminated. Thus, the protocol provides a detailed account of activities that the experimenter will face. For example, protocols often include a practical timetable; specific guides of actions; details of the parties who will be involved in the study; comprehensive definitions of the intervention, data and measurements and the overall experimental procedures and materials.

The protocol generally follows a master template. These checklists often include established key components that any test should consider. For example, *clinicaltrials.gov*, provided by the US National Institutes of Health, invites researchers to register their trials ahead of time, and requires information on a set of 13 study elements (https://register.clinicaltrials.gov):

1 Study identification
2 Study status
3 Sponsor/collaborators
4 Oversight
5 Study description
6 Conditions
7 Study design
8 Arms/groups and interventions
9 Outcome measures
10 Eligibility
11 Contacts/locations,
12 Sharing statement
13 References

These sets of rules not only focus the experimenter on the salient issues they ought to take into account but also allow the scientific community an opportunity to systematically and methodically assess how the test was executed, and the validity of the causal estimates that it proposes. In part, this additional layer of peer review is made possible by the requirement to publish the experimental protocol; experiments are less likely to 'bury' the study should they not like the results or to 'fiddle' with the figures. They are also less likely to 'go fishing' for statistically significant results, as we discuss below. Overall, this creates transparency, accountability, safety as well as consistency and efficiency in science.

A closer look at the benefits of protocols as a quality control tool

Considering the benefits of experimental protocols in more detail, the US Food and Drug Administration (FDA, 2018) discusses how this instrument can advance good practice in evaluation research. The FDA is responsible for protecting public health by ensuring the safety, efficacy and security of foods, drugs and other related products. Among other tools, it does so by providing oversight of the procedures through which new products and interventions are approved for future consumption. Experiments form part of these procedures and amongst them the supervision and approval of new tests.

The FDA's guidelines suggest that having a protocol serves four main purposes. First, using protocols introduces *uniformity* in research practices. Unlike qualitative research and observational research strategies, which are characterised by more freedom in the way in which a study is administered, evaluative research necessitates a more formal framework. Once a research protocol is used for evaluative research, then not only the ordering of the content is controlled but the terminology, key terms and methods are also homogenised and standardised, as much as possible. This uniformity is welcomed, as the research process and its final report are not prosaic and should not be left to creative interpretation. Whether a requirement set out by the FDA should be an industry standard is arguable, but it highlights a growing recognition of the need for uniform practices.

Similarly, protocols are particularly pivotal for the purpose of *replications* in science. When they include all the necessary information for obtaining consistent results, protocols are akin to a 'cookbook', in which all the processes and actions that form part of the experiment are detailed. Freedman et al. (2017), as well as Baker (2016), have convincingly shown that adequate and comprehensive reporting in the protocol facilitates reproducibility (Casadevall & Fang, 2010; Festing & Altman, 2002; see additional practical stages in Drover & Ariel, 2015; Harrison & List, 2004; Leeuw & Schmeets, 2016; List, 2011; Welsh et al., 2013).

The second utility of a protocol to which we alluded earlier is that it drives the scholar to think about the adverse effects of the intervention. Importantly, the protocol can be viewed as a **pre-mortem diagnosis** or **pre-mortem analysis**: a consideration of all the factors that can potentially go wrong and ways to mitigate these concerns. This provides the context for scrutiny into potential problems and their solutions, with a balanced emphasis on implementation as well as safety and ethics. After all, field experiments involve human beings, so experimenters must consider the potential adverse effects on the participants and minimise them as much as possible. As such, this is relevant to all studies involving human participants, and institutional review boards (IRBs) and ethics committees have a long tradition of looking precisely at these questions: Is the welfare of the participants meticulously considered? Is there a potential for a backfire effect and, if so, what can be done to minimise this risk? Are there any potential risks to the researchers themselves? These are issues that, once introduced in the experimental protocol, are more likely to be considered (see more broadly in Neyroud, 2016).

A third and crucial purpose of the protocol is to reduce the number of *revisions* to the experiment. Indeed, changes in protocols are inevitable. There are natural variations over time – for example, availability or definition of data, randomisation procedures, change in treatment providers, revisions to the eligibility criteria, definition of the treatment and so on. Before the final version of the protocol is approved and signed off on, multiple iterations are commonplace. However, the researcher must be aware that each revision is expensive and will result in delays in the trial. More importantly, revisions may result in increasing the risk of harm to the persons being studied or adding to the nuisance of being studied (see review in Meinert, 2012, pp. 195–204). Therefore, revisions should be kept to a minimum whenever possible. The crucial recommendation, however, is to keep track of the changes and transparently report on these revisions.

The final reason – arguably the most important for an experimental protocol – is to force more *complete and full reporting*. To place this rationale in perspective, consider the following finding reported by the author of *Bad Pharma: How Drug Companies Mislead Doctors and Harm Patients*, Ben Goldcare (2014)[1]:

> Overall, for the treatments that we currently use today, the chances of a trial being published are around 50 percent. The trials with positive results are about twice [as likely] to be published as trials with negative results. So, we're missing half of the evidence that we're supposed to be using to make informed decisions. [And] we're not just missing any old half, we're selectively missing the unflattering half.

We note that going 'fishing for statistical significance', 'data dredging' or '*p*-hacking' are concerning practices, especially when it comes to evaluative research with real policy implications (see discussions by Head et al., 2015; Payne, 1974; Weisburd & Britt, 2014; Wicherts et al., 2016). Here, the researcher misuses data analysis in order to locate statistically significant patterns because these outcomes are more publishable. A common way of detecting statistically significant outcomes outside the parameters of the main effects is by conducting multiple tests of significance on selective subgroups of participants, by combining variables, omitting uncomfortable participants, altering the follow-up periods or breaking down the treatment components – until one or more of the statistical tests yield a finding that is significant under the usual $p = 0.05$ level (Gelman & Loken, 2013). Since any data set with a degree of randomness is likely to contain relationships, experimenters may use one or more of these techniques to celebrate a statistically significant outcome, which again is more publishable.

Thus, there is an inherent and systematic concern that published evidence is one-sided: only showing 'successful' interventions and established cause-and-effect relationships. What does not 'work' or is not statistically significant is less likely to be published, and therefore remains in the 'grey literature' that is more difficult to find. Indeed, the issue of selective reporting and what is referred to as 'selection bias'

---

[1]See interview in *Time* magazine (28 February 2013): https://healthland.time.com/2013/02/28/how-drug-companies-distort-science-qa-with-ben-goldacre/

causes only statistically significant results or extraordinary effect sizes to be published in leading journals, while non-significant results are often buried or delayed. This 'file-drawer problem' – the systematic bias that comes from not reporting tests in which no statistically significant differences were found, or studies in which differences were found in the opposite direction from the anticipated or desired effect – is a major concern in all sciences. Rosenthal (1979), Scargle (1999) and others have shown that the probability that a study is published depends on the statistical significance of its results. It is a source of concern because it drives policymakers to consider only interventions that are deemed effective and results in much less attention to those that are ineffective or potentially backfire.

The literature then misestimates treatments (especially when results are synthesised in meta-analyses), and it also misinforms the scientific community that may attempt to unknowingly replicate these studies, their possible mistakes or even harm that they may cause to participants. As we discussed in Chapter 1, science is an evolutionary process, with multiple building blocks of knowledge that form our understanding about causation. Publication of selective results creates an unclear or even misleading depiction of phenomena.

Experimental protocols can assist in combating these issues. They create a public registry of tests to be administered (e.g. https://prsinfo.clinicaltrials.gov). Registries of protocols place the onus on the researcher to publish the results of the experiment, no matter the results. Similarly, the researcher is less likely to deviate from the experimental plan when the results are not appealing and might otherwise feel pressure to fish for statistically significant findings. The publication of the protocol, therefore, exerts pressure on the experimenter to avoid deviating from the plan and, subsequently, to publish the outcomes regardless of their direction and overall effect.

What should be included in the experimental protocol?

Rule 58.120 of the FDA[2] provides a list of factors that a sound protocol should contain, and we introduce them below verbatim. These elements form the most basic list of requirements that any experiment should consider, before even the first case is randomly assigned into the study arms. Over the years, this list has expanded, with additional requirements about the disclosing of funders, relationships between the treatment provider and the evaluation team and elements associated with the

---

[2]The FDA Code of Federal Regulations is a codification of the general and permanent rules published in the Federal Register by the executive departments and agencies of the Federal Government. Title 21 [Revised as of 1 April 2019] includes a 'Good Laboratory Practice For Nonclinical Laboratory Studies', and under it, in subpart G, there are details of the basics of a 'Protocol for and Conduct of a Nonclinical Laboratory Study' in 'Sec. 58.120 Protocol'.

description of the treatment and comparison conditions. However, the overall framework and the minimum set of details that any reasonable protocol should include have remained the same:

1   Descriptive title and statement of the purpose of the study
2   Identification of the test and control units
3   The name of the sponsor
4   The address of the testing facility
5   Participants' details
6   Description of experimental design and methods to control for biases
7   Description of interventions and potential adverse effects
8   Type and frequency of tests, analyses and measurements
9   The records to be maintained
10 A statement of the proposed statistical methods to be used

Such protocols provide the impetus for scholars around the globe to develop protocols that are more closely linked to their own disciplines (e.g. Cameli et al., 2018). While experimental designs are meant to be universal and follow certain research architectures, there are still differences between the disciplines in the ways in which these designs are instituted. Therefore, there are different protocols. For example, the recruitment of cases in psychology laboratory experiments can look very different from the recruitment of participants for law enforcement experiments in field settings. Issues such as consent, blinding or double blinding and incentives paid to participants for taking part in the study, for example, are inherently different between research fields. Similarly, the question of sample size looks very different in experiments that analyse places as the units of analysis rather than individuals (as we reviewed in Chapter 3). Therefore, bespoke protocols for different research settings are welcomed.

One example is the protocol template called SPIRIT (Standard Protocol Items: Recommendation for Interventional Trials; see www.spirit-statement.org/trial-protocol-template). SPIRIT is a protocol based on widely endorsed industry standards and the accepted protocol template for the journal *Trials*. We recommend having a look at this tool.

Another detailed and practical template is the CrimPORT (Criminological Protocol for Operating Randomised Trials; Sherman & Strang, 2009). CrimPORT focuses on both the internal mechanisms of trials as well as the 'managerial elements involved in making the experiment happen' (Sherman & Strang, 2012, p. 402). The 12 sections of this protocol call on the researcher, prior to the administration of the experiment, to consider the fundamental definitions of the various factors linked to the study (e.g. the treatment, the measurement, the sample and other internal elements of the test). By paying close attention to the operational and practical details of the up-and-coming experiment, the CrimPORT leads to a pre-mortem diagnosis of the possible pitfalls the study may encounter and to find ways to remedy them. This includes the organisational dynamics that any trial entails, which are often overlooked when planning experiments. For ease of reference, the protocol is provided in the appendix to this chapter, and the core factors are listed below:

1 Name [of experiment] and hypotheses
2 Organisational framework
3 Unit of analysis
4 Eligibility criteria
5 Pipeline: recruitment or extraction of cases
6 Timing
7 Random assignment
8 Treatment and comparison of elements
9 Measuring and managing treatments
10 Measuring and monitoring outcomes
11 Analysis plan
12 Due date and dissemination plan

Box 5.1

The Experimental Protocol

Experimenters will find the structure of the CrimPORT useful, and indeed a growing list of trials have utilised it, including many of those we have discussed in this book as illustrations of experiments in criminology. For example, the Rialto Police Department wearable cameras experiment experimental protocol (Ariel & Farrar, 2012), Operation turning point: an experiment in 'offender-desistance policing' (Neyroud, 2011; on implementation challenges in this trial, see Neyroud & Slothower, 2015), Restorative Justice Elements versus the Duluth model for domestic violence (Mills et al., 2012), the Salt Lake City court-mandated restorative justice treatment for domestic batterers experiment (Mills et al., 2015a, 2015b), Restorative Justice Conferences in Denmark (Van Mastrigt et al., 2018) and the Sycamore Tree victim awareness programme for released prisoners (Wilson, 2013).

The elements of the CrimPORT are generally intuitive and lay out the framework of the necessary conditions for successful trials. At the same time, we note that some articles of the CrimPORT can be complicated due to the nature of the issues they raise. As an illustration of the difficulties these protocols highlight, consider the first item of the CrimPORT: 'defining the hypotheses of the experiment'. The CrimPORT logically purports that any experiment requires a hypothesis that specifies an anticipated causal relationship between the independent and the dependent variables (or no relationship, as in the null hypothesis), so that statistical tests of significance and measures of effect size can be performed (for a more detailed review, see Kendall, 2003). Experimental hypotheses ought to be quantifiable, specific and formulated in advance of the experiment. The specificity and accuracy of the empirical research question are vital, and therefore great attention should be given to its conceptual and operational definition. What stimulus is being tested, *precisely*? To what counterfactual conditions or alternative treatments is the treatment effect compared? What are the conditions under which the null hypothesis will be falsified? These can be difficult questions to answer.

One element that is not required by the CrimPORT is the need for a literature review. The APA, which publishes guidelines on how to write academic papers and is considered by many to be an authoritative guideline for these matters, requests a brief literature review to support a study's hypothesis. A reasonable literature review must summarise the state of available knowledge as well as the systematic process that was used to arrive at this literature review (see Baumeister & Leary, 1997). The literature review should incorporate a summary of findings from systematic reviews and meta-analyses, if there are any, as they should form the strongest basis for laying out the hypotheses of the experiment.

Box 5.2

Synthesising Experimental Evidence Using Meta-Analyses

Meta-analysis as a research method is well recognised and used extensively in the social sciences (Lipsey & Wilson, 1993; Petrosino, 1995). It is a method of synthesising evidence on the same research question. Its value, as Petrosino (1995) persuasively argued, is that it supports policy, programme decision-makers and practitioners 'looking for more conclusive evidence before taking action may find a meta-analysis of many studies bearing on the issue more persuasive than a single one' (p. 275). The term *meta-analysis* was coined by Glass (1976) to refer to the 'statistical analysis of a large collection of analysis results from individual studies for the purpose of integrating the findings' (p. 3). Card (2015) later described it as 'a form of research synthesis in which conclusions are based on the statistical analysis of effect sizes from individual studies', which provide a 'statistically defensible approach to synthesising empirical findings' (pp. 7–8). Hunter and Schmidt (2004) were as clear, however less technical, in their articulation that meta-analysis is a 'process of cleaning up and making sense of research literatures that not only reveals the cumulative knowledge that is there, however also provides clearer directions about what the remaining research needs are' (p. 21).

We emphasise that the literature review requirement does not mean that the experiment must endorse a particular theory, as a theoretical framework is not a necessary condition for hypothesis testing. For experiments to reliably and validly demonstrate cause-and-effect relationships, we do not require a 'formal statement of the rules on which a subject of study is based, or of ideas that are suggested to explain a fact or ... an opinion' ('theory', Cambridge Dictionary). The strength of a test is not its ability to falsify a theory but rather to falsify the specific and quantifiable *null* hypothesis that is put to the test within the experimental model. The results of experiments can be problematised within theories, but a theory per se is not a *sine qua non* for conducting an experiment. Hypothesis testing involves a logical expression about the causal mechanism. Necessitating a theory for *every* test would lead us to ignore inductive (as opposed to deductive) probable causes (Johnson, 1932).

It is worth noting that experiments are in a position to falsify the null hypothesis, but not to prove the research hypothesis. Imagine an experiment that is perfectly designed and executed, with truly reliable data and appropriate control for all threats to validity, but which finds no statistically significant differences between treatment and non-treatment. One would be quick to conclude that this result is a proof that the mechanism has failed, as the treatment failed to produce the hypothesised effect. However, failing to reject the null hypothesis indicates that our sample did not provide sufficient evidence to conclude that there is a consequential relationship between the independent and the dependent variables. This lack of evidence does not prove that the effect does not exist. The null hypothesis is assumed to be the status of the relationship between the two variables, until contrary evidence proves otherwise. Nevertheless, the new evidence does not mean that the null hypothesis is true, but it allows us to conclude that the null hypothesis has not been disproven. We think the analogy described often in the literature to convey the difference between 'failing to reject' the null hypothesis and 'accepting' the null hypothesis was best illustrated by Taylor (2019):

> The presumption at the outset of the trial [in a criminal case] is that the defendant is innocent. In theory, there is no need for the defendant to prove that he or she is innocent. The burden of proof is on the prosecuting attorney, who must marshal enough evidence to convince the jury that the defendant is guilty beyond a reasonable doubt. Likewise, in a test of significance, a scientist can only reject the null hypothesis by providing evidence for the alternative hypothesis. If there is not enough evidence in a trial to demonstrate guilt, then the defendant is declared 'not guilty'. This claim has nothing to do with innocence; it merely reflects the fact that the prosecution failed to provide enough evidence of guilt. In a similar way, a failure to reject the null hypothesis in a significance test does not mean that the null hypothesis is true. It only means that the scientist was unable to provide enough evidence for the alternative hypothesis.

Protocols for reporting the findings of the experiment

While the pre-experimental protocol such as the CrimPORT is useful for planning experiments, we need bespoke templates to report the findings as well. The need for better reporting is shown by multiple reviews of existing experiments that have found dramatic underreporting or misreporting of key information that would make the experiment reproducible. Moher et al. (2015) have shown that fewer than 20% of popular publications in the life sciences have adequate descriptions of study design and analytic methods. A.E. Perry et al. (2010) found even more concerning non-compliance with proper reporting rules in criminology, reaching the conclusion that 'the state of descriptive validity in crime and justice is inadequate' (p. 245). Thus, more 'accurate and comprehensive documentation for experimental activities is critical', remark Giraldo et al. (2018, p. 2), because 'knowing how the data were produced is . . . important'.

By using templates for sharing the evidence and research methods with the wider scientific community, we hope to introduce consistency, clarity, transparency and accountability to the process of disseminating the results of experiments. Reporting standards for experiments were therefore developed over the years. One of the most popular choices is the CONSORT (Consolidated Standards of Reporting Trials; see review in M.K. Campbell et al., 2004; Montgomery et al., 2018). Another is TIDieR – the Template for Intervention Description and Replication (Hoffmann et al., 2014), although the CONSORT is more prevalent.

CONSORT is a 25-item checklist that guides reporting of trials. It lists a set of recommendations for reporting the results of the experiment, using a standard way to prepare 'reports of trial findings, facilitating their complete and transparent reporting, and aiding their critical appraisal and interpretation' (www.consort-statement.org). For example, the checklist items focus on reporting how the trial was designed, analysed and interpreted. Perhaps the most obvious point in the CONSORT checklist is to include 'randomised controlled trial' in the title of the publication. This is both to make clear what the paper is about and also to make it easier for others to find the publication (e.g. if conducting a systematic review or just looking for evidence on a given topic).

CONSORT also includes a flow diagram, which displays the progress of all participants through the trial. The usefulness of the flowchart cannot be overstated and should form part of any report, tracking trial participants from the point of recruitment or randomisation through to outcome reporting and analysis. This is where rigour in terms of 'attention to detail' comes in – we need to be able to track and report on all the individuals included in the trial from the point they were randomised through to our analysis.

In addition, extensions of the CONSORT statement have been developed to give additional guidance for RCTs with specific designs, data and interventions. There are several versions of CONSORT, tailored to minimum reporting requirements for a range of trial scenarios (e.g. individually randomised and cluster-randomised trials), including a new set of guidelines for reporting social interventions (Montgomery et al., 2018; see also www.consort-statement.org) or complex and multicentre experiments (Taxman & Rhodes, 2010).

Box 5.3

CONSORT Reporting Domains and Flowchart

Table 5.1 provides the CONSORT reporting domains and Figure 5.1 provides the CONSORT flowchart (www.consort-statement.org/consort-statement/flow-diagram)

Table 5.1 CONSORT reporting domains

CONSORT Reporting Domains	Item Number
Title and abstract[a]	1a, 1b
Introduction: background	2a
Introduction: objectives [a]	2b
Methods: trial design [a]	3a, 3b
Methods: participants [a]	4a, 4b
Methods: interventions [a]	5, 5a, 5b, 5c
Methods: outcomes	6a, 6b
Methods: sample size	7a, 7b
Randomisation: sequence generation	8a, 8b
Randomisation: allocation concealment mechanism	9
Randomisation: implementation	10
Randomisation: awareness of assignment	11a, 11b
Randomisation: analytical methods [a]	12a, 12b
Results: participant flow [a]	13a, 13b
Results: recruitment	14a, 14b
Results: baseline data [a]	15
Results: numbers analysed	16
Results: outcomes and estimation [a]	17a, 17b
Results: ancillary analyses	18
Results: harms	19

CONSORT Reporting Domains	Item Number
Discussion: limitations	20
Discussion: generalisability	21
Discussion: interpretation	22
Important information…	
Registration	23
Protocol	24
Declaration of interests[a]	25
Stakeholder involvement	26a, 26b, 26c

Note. CONSORT = The Consolidated Standards of Reporting Trials.

[a]Updated in Montgomery et al. (2018).

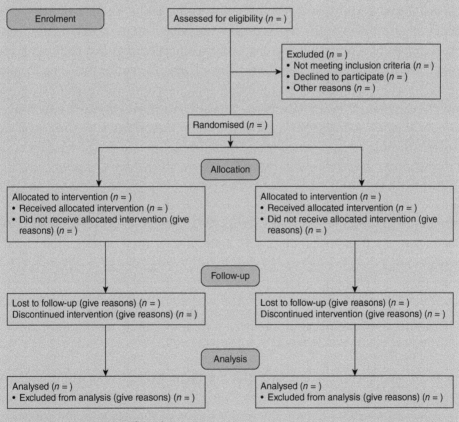

Figure 5.1 CONSORT flowchart

Note. CONSORT = The Consolidated Standards of Reporting Trials.

Implementation of experiments

For the most part, issues associated with 'how to' conduct experiments have been left undiscussed in mainstream experimental criminology, largely because they are not necessarily 'interesting' as assessed by leading peer-reviewed journals. However, the management, administration and politics of experiments are controversially complex and multifaceted. In these frameworks, one can locate the 'craft' of running experiments, and at least three steps are crucial: (1) the creation of a coalition for the purpose of the experiment, (2) the incorporation of field managers and *pracademics* in field experiments and (3) intensifying our focus on implementation sciences. Let us now consider these implementation elements in greater detail.

Researcher–practitioner coalitions

In experiments, particularly field trials, where the experimenter must rely on a treatment provider to deliver the intervention – the police, courts, charities or schools – a special type of relationship emerges between the researcher and the practitioner. Unlike medicine, where relationships between research universities and hospitals are well established, we have yet to find such an intertwined network of academics and practitioners in the social sciences. Most experimental projects are short-lived, purpose-oriented and depend on the actors rather than systems for their longevity. Whether the interest in conducting an experiment arises from the researcher or from the practitioner, collaboration is required throughout the entire experimental cycle (Garner & Visher, 2003). This cooperative process requires the intimate and continuous involvement of all sides – a conclusion reached in a wide range of research settings (Braga & Hinkle, 2010; Feder et al., 2011; J.R. Greene, 2010; Sherman, 2015; Sherman et al., 2014; Weisburd, 2005). Reflecting on these studies, Strang (2012) concluded that

> experiments require close cooperation between the parties because of the need for maintenance and monitoring. . . . Relationships which may be characterised as *temporary coalitions* [emphasis added] for a common purpose may, under the right conditions, ultimately mature into true research partnership. (p. 211)

Therefore, we concur that the successful organisational framework in experiments is one that is characterised by a *coalition* of temporary interests that require a particular kind of leadership. Indeed, the establishment of the relationship between the research team and the staff involved in programme delivery (both leadership as well as operational staff) is the first step in setting up an experiment, followed by a contractual and organisational framework. However, it seems that considerable attention must be given by this coalition to the ongoing cooperation. The day-to-day pressures of running a major project can hinder the success of the experiment, whether a legal memorandum of understanding and willingness on behalf of the leadership is present or not. Operational staff must be invested, both emotionally and professionally, in the success of the intervention; otherwise, the experiment is unlikely to succeed. This is particularly the case when there is lack of respect between field staff members and headquarters leaders (as seems to be the case in Israel, for both the police and the national education system; see Brants-Sabo & Ariel, 2020; Jonathan-Zamir et al., 2019).

Meaningful experiments often disrupt the daily routines of treatment providers, and consequently, the programme of change requires active agreement to comply with the research protocol. Otherwise, a host of problems can 'go wrong', ranging from getting cases into the experiment, screening for eligibility, managing random assignment, attrition from the study, consistency of delivery of the experimental conditions and monitoring and measuring programme delivery (Strang, 2012, pp. 217–222). Therefore, a strong and ongoing collaborative approach is needed to enhance the likelihood that the experiment will be delivered with integrity (Weisburd, 2000). Mutual consent, emotional investment and genuine belief in the purpose of the experiment are required; otherwise, the experiment is less likely to succeed (Boruch, 1997).

In brief, we note that in clinical trials there are five phases – but we can extrapolate from these guidelines for field tests in the social sciences as well: the earlier phases look at whether an intervention is safe or has side effects, while later phases test the effectiveness of the intervention versus control conditions. Phase 0 and phase 1 are small trials with usually between 10 and 50 participants, which aim to test the harmfulness of the intervention, without comparison groups. Phase 2 may or may not have randomised allocation of participants but would have a comparison group to test the effect of the intervention on a sample of 100 against control conditions. Phase 3 refers to larger trials, which usually use random assignment of hundreds or thousands of participants – the common experiments we discussed in this book. Phase 4 usually refers to longitudinal studies to investigate long-term benefits and side effects. For more details, see Cancer Research UK (2019).

The necessary role of field managers and the emergence of the pracademic

In recent years, a lot has been written about the role of '**pracademics**' in research (Braga, 2016; Huey & Mitchell, 2016; Magnusson, 2020; McCabe et al., 2016; Piza et al., 2020; Sherman, 2013; Volpe & Chandler, 2001; Willis, 2016). The term refers to practitioner-academics: those who span the theoretical world of academia as a scholar and the pragmatic world of practice (Walker, 2010; Walker et al., 2008). These are individuals with sufficient experience in the field to appreciate the nuances and subtle signals that the environment emits, and also possess academic research skills to systematically assess patterns, probe deeply into causal issues and understand implications for practice.

Panda (2014) argues that the function of the 'pracademic' is to actively question practice, as they usually have gained skills in critically observing the world and have developed a repertoire of approaches to seeing 'reality' with varying levels of analysis. These are either practitioners who become engaged through their association with academics via a research project or a broader research agenda or academics hired by the treatment agency with the aim of transforming the department into an evidence-led organisation. An exemplar of the former is Commander Alex Murray of the Metropolitan Police in London, who has established the Society of Evidence Based Policing, and has one 'foot in academia', playing a key role in many policing RCTs while managing one of the busiest portfolios in British policing (e.g. Johnson et al., 2017; A. Murray, 2013). Dr Geoffrey Barnes is an example of the latter, a career academic who has been appointed to the role of embedded director of criminology in Perth, Australia, and then in London, UK.

Experience has taught us that without the direct and immediate involvement of a pracademic, there is a much lower chance that field experiments can be successfully administered. The role of the field research manager, who can continually monitor the state of play in both the research team as well as the practitioner team, and between the teams (Strang, 2012), is vital. In this sense, the crucial defining feature of a successful pracademic is their stake in the successful administration of the trial, across both practice and academia. The field manager cannot be remunerated above and beyond a usual wage for their time – either as a practitioner or a hired research assistant – so another type of incentive is required to create intrinsic engagement. Creating a research programme for the pracademic, where they could make an original intellectual contribution while also filling a managerial role for the experiment for research purposes, can be invaluable. A practitioner who undertakes an academic qualification as part of the experiment – a master's thesis or preferably a doctoral dissertation – will not only gain the necessary skills for *future* rigorous research through a proper mentoring programme (Braga et al., 2014; Eby et al., 2008) but would also have the emotional investment to help administer the experiment.

Finally, successful experiments depend on the ability of the pracademic to exert influence on their fellow practitioners – what Fixsen et al. (2005) refer to as 'purveyors'. Otherwise, the experiment may not materialise (Sykes, 2015). It is important that the pracademic field manager has access to data, the personnel and the social capital required to mobilise the agency, vertically and horizontally, for the purpose of the experimental collation. While we cannot offer causal evidence to that effect, observational data suggests that the pracademic route is often successful (Ariel, Garner, et al., 2019).

Mainstreaming implementation science

While experimental work has developed and aided in identifying valid causal estimates of evidence-based interventions, the science related to implementing experiments with fidelity lags far behind. 'Implementation' in this sense means the efforts to incorporate a process of intervention (as opposed to its outcomes), and the literature in this matter, across domains, is limited. Fixsen et al. (2005) were able to synthesise 377 published studies in the area of implementation that included, among others, experimental evaluations of implementation factors that would be viewed as 'successful' (which amounted to only 22 out of the 377 studies). Based on this review of the available evidence, the authors identified a certain set of conditions that, when present, appear to contribute to successful implementation:

(a) carefully selected practitioners receive coordinated training, coaching, and frequent performance assessments;

(b) organizations provide the infrastructure necessary for timely training, skilful supervision and coaching, and regular process and outcome evaluations;

(c) communities and consumers are fully involved in the selection and evaluation of programs and practices; and

(d) state and federal funding avenues, policies, and regulations create a hospitable environment for implementation and program operations. (p. vi.)

The evidence on what can be viewed as successful implementations of experimental programmes appears ubiquitous (as the reviewers have not limited the review to a particular field). However, the operationalisation of 'implementation' is not straightforward because implementation does not usually follow a logically progressing manual, as we would expect to see in statistical analyses, engineering or medical procedures. The latter are fairly established procedures, and formal guidelines on conducting tests of significance – or diagnosing and subsequently treating patients – are largely scripted. These guidelines are of course no recipe for guaranteed success, but they provide information for future researchers. The same cannot be said about the implementation of successful experiments because many issues with implementation are localised, political and subjective. The above-mentioned guidelines are useful, but remain abstract, and our knowledge on how to manufacture experiments with rigour remains inherently subjective. We therefore need more implementation research, in different areas, in order to create more established practices.

On the other hand, one feature that is consistently crucial in any experiment is the *tracking* of inputs, outputs and data outcomes (see Sherman, 2013). While the concept of tracking has been used in evidence-based policing more broadly as a line of research (see e.g. Damen, 2017; De Brito & Ariel, 2017; Dulachan, 2014; Gresswell, 2018; Henderson, 2014; Jenkins, 2018; Pegram, 2016; Rowlinson, 2015; Young, 2014), we can generalise from this body of evidence two key lessons for experimental designs more broadly, which can help mainstream implementation sciences more robustly.

First, experimenters must keep a detailed account of data – and not just about the experiment but also about the ways in which experiment was implemented: how many meetings were held, with whom and under what conditions? What is the composition of the research team, and how and what was done by each member to execute the research plan? What is the process of endorsement of the research within the treatment-delivery team? How many hours were dedicated for each element of the experimental cycle? What mechanisms were placed to assure buy-in from stakeholders? These and other relevant tracking questions about the implementation of the experiment, from an organisational and structural side, are crucial if we want to be able to assess the application of the test and the fidelity of the experiment. Without tracking data, we may be left without an answer about the mechanisms that lead to the causal estimates found in the study (e.g. Grossmith et al., 2015).

Second, implementation queries are not just important for process evaluations but also critical from an external validity perspective. As we reviewed in Chapter 3, external validity is always a concern in experimental research, as many argue that, at least, in field settings an experiment does not mimic real-life conditions but rather provides a synthetic appearance of true population means. In part, this argument has weight because we are unable to quantify the environmental and contextual settings of the experiment due to a lack of observable data. Implementation science comes to correct for this bias by requiring a detailed qualitative account of the experiment. The more details shared, the more we can learn about the generalisability of the study's outcomes.

Ethics in experiments

The issue of ethics is not currently covered in the CrimPORT template, but we sense that it should be included, as it is part of the experimental architecture (e.g. Stern & Lomax, 1997, and as introduced in *The SAGE Quantitative Research Kit*, Volume 1), especially within those studies involving human participants. Concerns about the ethical viability of randomised experiments have been consistently cited as one of the main impediments to their proliferation (Clarke & Cornish, 1972; Neyroud, 2017). While ethical considerations may come in a variety of shapes and sizes, there are two prevalent themes of ethical concerns. One is a perceived lack of fairness in denying treatments to control group participants. A second is the perception of risk of harm to control group participants and the subsequent implications for the legitimacy of the agency. These themes unintentionally ignore fundamental points about random assignment – that there is an equal chance of receiving the treatment for all participants and that the benefit of treatment is usually unquantified at the point of experimentation. Yet they reflect a realpolitik concern present in many agencies and merit the attention of experimenters at the earliest stages of their design processes, principally in two respects: (1) securing the support of stakeholders and participants and (2) ensuring that the participants are as safe as possible.

Social scientists can draw ethical lessons from experiments in clinical trials (Boruch, Victor, et al., 2000; Weisburd, 2005). In medical trials, the dominant fundamental ethos is taken from the Hippocratic Oath often summarised as 'first, do no harm'.[3] If participants are at risk, then there must be a good justification for potentially harming them. Experimentation may be considered *necessary* only when there is 'collective equipoise', where the medical community is genuinely uncertain about the effects of a desired treatment. Under these conditions, tests are needed to make sure that future patients will not be harmed as a result of the tested treatment. There is always at least a marginal risk inherent in any situation where a participant takes part in a study, and if the study places an individual at risk, it had better be for a good reason.

This ethical synopsis in medicine is reasonable, as one would expect to result from the maturity of clinical trial practice in medical research. However, social science–based trials are considerably less mature and experienced than biomedical trials (see the Global Policing Database by Mazerolle et al., 2017; as well as Higginson et al., 2014). We do not have enough experience in terms of the ethical considerations. This lack of experience leads many to reach the wrong conclusions about the need to conduct more experiments, not less (see review in Neyroud, 2016; see also Mark & Lenz-Watson, 2011).

One example is randomisation. Contrary to what antagonists of RCTs think, the random allocation of participants into the study arms is *more* ethical than any other treatment allocation procedure. If the study aims to test the effectiveness of a treatment, and the researcher *knows* that this particular treatment is indeed effective, then *not* allocating this treatment to patients may be seen as unethical. After all, the main

---

[3]For this reason, the World Health Organization stopped an experiment on efficacy and safety of hydroxychloroquine and azithromycin for the treatment of ambulatory patients with mild COVID-19, as it was found during the trial that the drug presents serious health risks to those who take it (see story on *BBC*, 22 May 2020; www.bbc.co.uk/news/world-52779309). The evidence did not deter President Donald Trump from publicly endorsing it and claiming to have taken it to prevent COVID-19 (which, as we know, did not help him from catching and possibly superspreading the virus). As in other areas of his presidency, Trump epitomised an evidence-free policymaking approach (see Drezner, 2020).

purpose of science is to discover how to reduce harm and promote well-being. If the science behind the treatment has been 'proven' to the point that we know that the treatment 'works', we must give it to those who can benefit from it. *However*, the issue is that we often do not know that the treatment is effective; we *assume* that it should, and therefore we conclude that randomly excluding participants from the treatment is unethical. Yet how do we know that it should without a test? Something that is logical is not necessarily true, nor does the fact that a treatment has been practised for a long time (prisons, arrests, fines and counselling) mean that it has a desirable effect. If you *know* that the intervention positively affects well-being, then flipping a coin that would prevent treatment is indeed unfair. When you do not know – and hence the reason for conducting an experiment in the first place – then dividing the risk of a backfire effect is the fairest when it is done randomly.

A template for a general assessment of ethics in experimental designs

We suggest a simple, three-part ethics framework for criminal justice researchers to consult when designing experimental protocols. The case for considering this framework is not to attract political support or financial sponsorship for an experimental design, although either or both may have a proxy benefit of giving attention to ethical concerns in a structured way. Our framework is instead aimed at an adapted version of the principle of 'first, do no harm'. In criminal justice, adaptation of this principle is required to allow for the variety of possible 'patients', which may feasibly include not just victims of crime but also offenders, police personnel, local community members and taxpayers, depending on the specific context of the experiment. The purpose of our proposed framework is to ensure no harm comes to any or all of these groups, which we suggest may be best achieved through consistent and explicit assessment. While the framework is not exhaustive, completion of an assessment such as this should help switch the focus of ethical debate around fairness and the risks of perceived 'denial of service' to inverse considerations about how to justify other less rigorous methods of evaluating effects (Weisburd, 2003).

Determination of informed consent

Consideration should be given to how informed consent can be sought from participants in a given experiment. This process will be completely dependent on the specific circumstances of the study, but seeking the agreement of people who take part in an experiment that benefits science but may hurt them is fundamental. Therefore, we want to make sure that they are participating freely and that they are willing to be exposed to known and unknown risks. Therefore, the rule and common practice is that every time an individual takes part in a study, their written consent should be required. IRB and ethics review boards take this issue very seriously.

However, there are exceptions to this rule. Rebers et al. (2016) have identified four exemptions to the rule of informed consent for research with an intervention: 'data validity and quality, major practical problems, distress or confusion of participants ... and privacy protection measures' (p. 9). For example, if an individual is legally required to take part in treatment, then consent is not possible or required, because there are no procedures for obtaining informed consent. For example, offender treatment programmes do not require consent from the offenders prior to assignment into treatment because the offender has no choice but to participate. Domestic abuser interventions are mandated by law in many states in the USA (see Mills et al., 2012), for example, and failure to take part in treatment results in punishment. There are many types of batterer intervention programmes that courts can mandate domestic offenders to attend (Babcock et al., 2004; Cheng et al., 2019), and the offenders do not get to pick them (unless there are compelling reasons). Consequently, an experiment that investigates the effects of such a programme cannot seek consent because it interferes with the natural processes of treatment. As the informative IRB rule dictated by the University of Michigan,[4] the requirement of informed consent can be waived when 'the research . . . involves no procedures for which written consent is normally required outside the research context'.

Furthermore, assignment to an experimental treatment can be seen as fair, even without obtaining informed consent, when the tested intervention is assumed to benefit the offender. If we hypothesise that the treatment has a harmful effect on the offender and that there are no benefits to anybody else from exposing the offender to the experimental treatment, then we should not conduct the experiment (in fact, we should not use it at all). The research hypothesis must therefore present a compelling reason why exposure to the stimulus would benefit the treatment participants, even if the term *benefit* were defined broadly, while consent is not sought.

The involvement of qualified researchers using appropriate research designs

Throughout this text, we have argued that designing experiments in field settings is a painstaking endeavour. Poorly designed experiments may lead to incorrect

---

[4]https://bit.ly/3pF3GHp

conclusions, and this may have negative consequences for participants, stakeholders, the wider research community and the general public, whose tax funds have probably paid for the experiment. It is important, then, that experimental designs are planned with the involvement of a suitably qualified designer.

We do not prescribe what such an appropriate qualification may be, but we suggest that completing an ethics assessment should address this issue and give specific regard to the experience, and independence, of the researchers involved. Devereaux et al. (2005) go as far as suggesting that 'increased use of the expertise based design will enhance the validity, applicability, feasibility, and ethical integrity of randomised controlled trials' (p. 388). At minimum, the qualified researcher must be well-versed in the experimental architecture, because a poorly executed study can then create unnecessary risk to the participants. If the experiment fails, then whatever the level of risk the stimulus and the participation in an experiment present to the participants cannot be justified. For example, if we plan an experiment that is underpowered (i.e. there are not enough participants to show a statistically significant effect), then exposure to a risky intervention would be deemed unnecessary and patently unfair. Knowledge in these and related matters is therefore an ethical requirement.

Assessment of potential benefits and risks

Experiments are complex, take time to complete and are often expensive. Experimenters should consider, therefore, the scale of the potential benefits of the intervention under examination and compare these to the potential risks involved. Such assessments may of course take many forms – ranging from the quantitative to the qualitative assessments; however, prescribing what form they should take is less important than the act of explicitly weighing benefits against risks at the outset of a design.

Conclusion

Experimental tools that promote the well-being of participants as well as the efficient operation of the experiment are important. Checklists, reviews and templates promote good scientific practice, as they enhance transparency, accountability and overall competence. As we covered in this chapter, protocols have many benefits. For example, they force the experimenter to follow best practice guidelines: a detailed and publicised blueprint limits the likelihood of going 'fishing for statistical significance'. Similarly, having a protocol implies that the researcher is obliged to publish the results of the experiment, analysed based on the parameters set forth in the experimental protocol. There are no strict rules about the publication format – in peer-reviewed journals, on academic websites[5] or at academic conferences (although we generally preferred the peer-review route, like the journal *Trials*,[6] because it provides another layer of control and oversight to the scientific inquiry). Either way, the protocol usually incorporates a section that stipulates that a publication is an integral element of the research. Thus, there is a greater likelihood that experimenters who use a protocol will also avoid the file-drawer bias.

We note that these issues are less concerning in observational research. For example, data mining is generally a good idea in big data studies, because emerging patterns and potential causal hypotheses can then be established for future research. Accidental discoveries are usually detected this way – for example, that arrests in domestic abuse cases deter employed offenders, but they appear to increase recidivism for unemployed offenders with fewer stakes in conformity (Ariel & Sherman, 2012; Sherman et al., 1992; Toby, 1957). On the other hand, in the context of experiments that seek to establish cause-and-effect relationships, the main concern is the primary effect of the independent variable on the dependent variable. A closer look at subgroups, categories of data and sections of the data must be declared ahead of time, and preferably as part of the overall experimental design (through blocking, stratification and minimisation, as described in Chapter 3) – and the forum where such ex ante declarations of data analysis and breakdown should be made is the experimental protocol.

We then made the case for using a template for reporting experiments. Using tools like the CONSORT statement and its flowchart introduces transparency, homogeneity and clarity in the reporting of experimental designs. The more researchers apply them, the higher the likelihood that we will create a seamless collection of evidence on any causal phenomenon. This may be particularly useful when trying to understand the value of the experimental results for participants such as those who were exposed to the intervention. However, there are clear advantages for using these post-experimental protocols for reporting the findings for future systematic reviews and meta-analyses.

The second portion of this chapter was dedicated to other necessary steps to follow when implementing experiments with rigour. We have reviewed the benefits of field managers and the pracademic for the advancement of science. Without a dedicated staff member who also has a footing in science, in some capacity, experiments are more challenging. A person with a vested interest to complete an advanced degree, for example, but who also has an official position as the treatment provider, can be the ideal field manager. We also argued that implementation sciences have much to

---

[5] For example, https://www.crim.cam.ac.uk/global/docs/crimport.pdf/view

[6] https://trialsjournal.biomedcentral.com/

EXPERIMENTAL DESIGNS

teach us, and the evidence gradually accumulated in the 'how to' sciences of experiments can be informative and helpful. Tracking of all the inputs, outputs and outcomes of the experiment can be useful as well, because such research informs us as to the conditions under which experiments can be more successfully implemented.

Finally, we discussed ethics in experiments. One element to highlight is our argument for the ethical consideration of randomisation: It generates a *fair* comparison group. Therefore, these should be exploited in situations when one treatment *may* be better than another treatment, but we want to *know* that with greater certainty. However, other benefits to randomisation make random assignment much more useful to researchers, practitioners and policymakers. Within this context, we suggested a three-tier process of considering the ethical dimensions of the experiment, which supplement the usual IRB ethics review processes. The more researchers consider issues such as the benefits and risks to participants, the consent needed from participants and what to do in lieu of informed consent, as well as the involvement of an experienced experimenter, the better it would be for proper scientific exploration.

Appendix: CrimPORT template

Contents

1. NAME AND HYPOTHESES
2. ORGANISATIONAL FRAMEWORK
3. UNIT OF ANALYSIS
4. ELIGIBILITY CRITERIA
5. PIPELINE: RECRUITMENT OR EXTRACTION OF CASES
6. TIMING
7. RANDOM ASSIGNMENT
8. TREATMENT AND COMPARISON ELEMENTS
9. MEASURING AND MANAGING TREATMENTS
10. MEASURING OUTCOMES
11. ANALYSIS PLAN
12. DUE DATE AND DISSEMINATION PLAN

1. Name and Hypotheses

1.1 Name of Experiment:

_____

1.2 Principal Investigator:

1.2.1 (Name) _____

1.2.2 (Employer) _____

1.3 <u>1st **Co-Principal Investigator:**</u>

    1.3.1 (Name) _____

    1.3.2 (Employer) _____

1.4 <u>**General Hypothesis**</u>:

(Experimental or Primary Treatment) _____ causes (less or more) _____ (crime or justice outcome) _____ than (comparison or control treatment) _____.

1.5 <u>**Specific Hypotheses:**</u>

    1.5.1 List all variations of treatment delivery to be tested.

    1.5.2 List all variations of outcome measures to be tested.

    1.5.3 List all subgroups to be tested for all varieties of outcome measures.

2. Organisational framework

(Check only one from a, b, c, or d)

    2.1 In-House delivery of treatments, data collection, and analysis

    2.2 Dual Partnership: Operating agency delivers treatments with independent research organisation providing random assignment, data collection, analysis

        2.2.1 Name of Operating Agency _____

        2.2.2 Name of Research Organisation _____

    2.3 Multi-Agency Partnership: Operating agencies deliver treatments with independent research organization providing random assignment, data collection, analysis

        2.3.1 Name of Operating Agency 1 _____

        2.3.2 Name of Operating Agency 2 _____

        2.3.3 Name of Operating Agency 3 _____

        2.3.4 Name of Research Organization _____

    2.4 Other Framework (describe in detail):

3. Unit of analysis

(Check only one)

    3.1 __A. People (describe role: offenders, victims, etc.) _____

    3.2 __B. Places (describe category: school, corner, face-block, etc.) _____

3.3 __C. Situations (describe category: police–citizen encounters, fights, etc.) _____

3.4 __D. Other (describe) _____

4. Eligibility criteria

4.1 Criteria Required (list all)
    4.1.1 . . .

4.2 Criteria for Exclusion (list all)
    4.2.1 . . .

5. Pipeline: Recruitment or extraction of cases

(Answer all questions)

5.1 Where will cases come from?

5.2 Who will obtain them?

5.3 How will they be identified?

5.4 How will each case be screened for eligibility?

5.5 Who will register the case identifiers prior to random assignment?

5.6 What social relationships must be maintained to keep cases coming into the experimental pipeline?

5.7 Has a Phase I (no control, 'dry-run') test of the pipeline and treatment process been conducted? If so,
    5.7.1 how many cases were attempted to be treated?
    5.7.2 how many treatments were successfully delivered?
    5.7.3 how many cases were lost during treatment delivery?

6. Timing: Cases come into the experiment according to

(Check only one)

6.1 A trickle flow process, one case at a time ___

6.2 A single batch assignment ___

6.3 Repeated batch assignments ___

6.4 Other (describe below) ___

7. Random assignment

7.1 How will a random assignment sequence be generated?

(Coin toss, every Nth case, and other non-random tools are banned from CCR-RCT).

(Check one from 1, 2, or 3 below)

> **7.1.1** Random numbers table case number sequence sealed envelopes with case numbers outside and treatment assignment inside, with two-sheet paper surrounding treatment __
>
> **7.1.2** Random numbers case-treatment generator programme on secure computer __
>
> **7.1.3** Other (please describe below) __

7.2 Who is entitled to issue random assignments of treatments?

> **7.2.1** Role:
>
> **7.2.2** Organisation:

7.3 How will random assignments be recorded in relation to case registration?

> **7.3.1** Name of database:
>
> **7.3.2** Location of data entry:
>
> **7.3.3** Persons performing data entry:

8. Treatment and comparison elements

8.1 Experimental or Primary Treatment

> **8.1.1** What elements must happen, with dosage level (if measured) indicated.
>
> > **8.1.1.1** Element A:
> >
> > **8.1.1.2** Element B:
> >
> > **8.1.1.3** Element C:
> >
> > **8.1.1.4** Other elements:

8.1.2 What elements must not happen, with dosage level (if measured) indicated.

> > **8.1.2.1** Element A:
> >
> > **8.1.2.2** Element B:
> >
> > **8.1.2.3** Element C:
> >
> > **8.1.2.4** Other elements:

8.2 Control or Secondary Comparison Treatment

> **8.2.1** What elements must happen, with dosage level (if measured) indicated.
>
> > **8.2.1.1** Element A:

8.2.1.2 Element B:

8.2.1.3 Element C:

8.2.1.4 Other elements:

8.2.2 What elements must not happen, with dosage level (if measured) indicated.

8.2.2.1 Element A:

8.2.2.2 Element B:

8.2.2.3 Element C:

8.2.2.4 Other elements:

9. Measuring and managing treatments

9.1 Measurements

9.1.1 How will treatments be measured?

9.1.2 Who will measure them?

9.1.3 How will data be collected?

9.1.4 How will data be stored?

9.1.5 Will data be audited?

9.1.6 If audited, who will do it?

9.1.7 How will data collection reliability be estimated?

9.1.8 Will data collection vary by treatment type?

If so, how? _____

9.2 Management

9.2.1 Who will see the treatment measurement data?

9.2.2 How often will treatment measures be circulated to key leaders?

9.2.3 If treatment integrity is challenged, whose responsibility is correction?

10. Measuring and monitoring outcomes

10.1 Measurements

10.1.1 How will outcomes be measured?

10.1.2 Who will measure them?

10.1.3 How will data be collected?

10.1.4 How will data be stored?

10.1.5 Will data be audited?

10.1.6 If audited, who will do it?

10.1.7 How will data collection reliability be estimated?

10.1.8 Will data collection vary by treatment type?

If so, how? _____

10.2 Monitoring

    10.2.1 How often will outcome data be monitored?

    10.2.2 Who will see the outcome monitoring data?

    10.2.3 When will outcome measures be circulated to key leaders?

    10.2.4 If experiment finds early significant differences, what procedure is to be followed?

11. Analysis plan

11.1 Which outcome measure is considered to be the primary indicator of a difference between experimental treatment and comparison group?

11.2 What is the minimum sample size to be used to analyse outcomes?

11.3 Will all analyses employ an intention-to-treat framework?

11.4 What is the threshold below which the percent of treatment-as-delivered would be so low as to bar any analysis of outcomes?

11.5 Who will do the data analysis?

11.6 What statistic will be used to estimate effect size?

11.7 What statistic will be used to calculate p values?

11.8 What is the magnitude of effect needed for a $p = .05$ difference to have an 80% chance of detection with the projected sample size (optional but recommended calculation of power curve) for the primary outcome measure?

12. Dissemination plan

12.1 What is the date by which the project agrees to file its first report on CCR-RCT? (Report of delay, preliminary findings, or final result).

12.2 Does the project agree to file an update every six months from date of first report until date of final report?

12.3 Will preliminary and final results be published, in a 250-word abstract, on CCR-RCT as soon as available?

12.4 Will CONSORT requirements be met in the final report for the project? (See www.consort-statement.org/)

12.5 What organisations will need to approve the final report? (Include any funders or sponsors).

12.6 Do all organisations involved agree that a final report shall be published after a maximum review period of six months from the principal investigator's certification of the report as final?

12.7 Does the principal investigator agree to post any changes in agreements affecting items 12.1 to 12.6 above?

12.8 Does the principal investigator agree to file a final report within two years of cessation of experimental operations, no matter what happened to the experiment? (e.g., 'random assignment broke down after three weeks and the experiment was cancelled' or 'only 15 cases were referred in the first 12 months and experiment was suspended').

Chapter Summary

- This chapter lays out the steps that experimenters need to consider when developing and managing field trials. We introduce three interconnected topics. First, we discuss the experimental protocol, a vehicle through which better research practices are made possible.
- There is the pre-experimental protocol, which is the blueprint of the research project. It is useful because writing up the detailed plan forces the researcher to conduct a 'pre-mortem diagnosis' of the possible pitfalls and risks the experiment may encounter – and then to offer solutions to remedy these threats.
- There is also the post-experimental protocol, usually in the form of a checklist, which addresses the reporting of the findings of the experiment. There are industry-standard templates that can help researchers report their findings and identify the key elements necessary for the scientific community to then be able to critically assess the study's findings. Importantly, these protocols are central for reproducibility purposes.
- We then discuss some of the more practical steps in the implementation of experiments in field settings. These operational issues need to be addressed if we want to produce the most valid causal estimates. Having the necessary organisational framework can 'make or break' experiments, no matter how potentially effective the tested treatment.
- We highlight the need for creating an efficient researcher–practitioner coalition, the necessary role of field manager and why we should pay attention to implementation science, which can inform us under which conditions the field experiment is more likely to succeed.
- We talk about the emergence of the pracademic as an entity who can streamline experiments in field settings, given their double role as both practitioners and academics.
- Finally, we discuss some of the ethical considerations that one must be aware of when conducting experiments. We make the case for ethical experimentation and suggest a framework for considering ethics in experimental designs.
- Our aim for this chapter is to provide a roadmap for anyone interested in designing and reporting on an effective trial.

Further Reading

United Nations Global Health Ethics Unit. (n.d.). *Ethical standards and procedures for research with human beings*. www.who.int/ethics/research/en
The issue of ethics in experimental research is so crucial that the United Nations dictated special guidelines on the principles to observe for the protection of research participants. This collection of documents, principles and standards is informative for any scholar considering applying experimental methods when human beings are involved. It also illustrates the global view on these ethical issues, indicating the universality of such ethics.

Schulz, K. F., Altman, D. G., Moher, D., & CONSORT Group. (2010). CONSORT 2010 statement: Updated guidelines for reporting parallel group randomised trials. *Trials, 11*(1), 32.

Turner, L., Shamseer, L., Altman, D. G., Schulz, K. F., & Moher, D. (2012). Does use of the CONSORT statement impact the completeness of reporting of randomised controlled trials published in medical journals? A Cochrane review. *Systematic Reviews, 1*(1), 60.

This chapter briefly introduced the benefits of research protocols for systematic reviews and meta-analyses. A reliable, consistent approach to reporting the findings from experiments is crucial for scholars when studying the overall body of evidence on one research question from multiple sources. Without systematic reporting of all relevant information about the study and its outcomes, future reviews of evidence may not accurately synthesise the data in meta-analyses of research. Schulz et al. offer a review of these issues and how protocols are in a position to solve such concerns; however, note their varying degrees of success, as discussed in this review by Turner et al.

6

CONCLUSIONS AND FINAL REMARKS

Chapter Overview

This book delved into the type of research method that is meant to uncover causal relationships between variables. There are many types of research strategies, but to answer questions that involve cause-and-effect expressions, experiments are required. There are different experimental designs at our disposal, and the fitness of any model is dependent on the contextual, theoretical and organisational factors that determine the choice of design.

In Chapter 1 as well as throughout this book, we have made the case for using RCTs over other types of causal designs when conducting evaluative research – particularly in impact evaluations involving policy and tactics. However, we are cognisant that other experimental designs exist, including advanced statistical designs that aim to create the conditions analogous to true experiments. These will remain popular and prevalent. Still, prospective, true experiments with random allocation of units are preferable over other experimental strategies.

Chapter 2 has paid considerable attention to the benefits of randomisation in causal research. The primary reason for the superiority of the random allocation of units into treatment and control arms of the study is its ability to create the most convincing counterfactual conditions. To talk about causality means using a comparison group, a reasonable benchmark. The ability of randomisation to create a comparable group through a fair and systematic process of allocating units into the study groups is supported by the fundamentals of mathematics and probability theory. Over time, with enough iterations (i.e. trials), the sample means would be as close to the true population means as possible. There are different methods of randomisation, including simple, block or minimisation approaches. Collectively, however, these processes create comparable groups to treatment groups, in all known and unknown, measured and unmeasured, factors, except that the treatment group is exposed to a stimulus, but the control group is not. These processes can be applied to different units of analysis, but, again, they are usually more advantageous than using non-randomisation techniques for creating pretreatment balance between the groups.

We then offered an exposition of the various threats that different research designs could face while making causal estimations. Chapter 3, which focused on the ways through which more control over the administration of the test can be applied, has highlighted the common threats to the validity of a test and how they can be mitigated. These hazards can be detrimental, and experimenters go to considerable lengths to convince the audience of their ability to control for these threats. With this in mind, however, researchers tend to agree that the threat of history, maturation, testing or instrumentation effects are serious concerns that may counter the validity of the study's conclusions. In a similar way, biases associated with regression to the mean, differential selection and treatment spillover effects remain a concern in every experimental design. As we reviewed the various risks, we arrived at the same conclusion reached by the 'classic' textbooks on research methods – Thomas D. Cook, Donald T. Campbell, Julian C. Stanley, Ronald A. Fisher and so on – that prospective, parallel groups assigned at random to treatment and control conditions provide the best defence against these risks. They do not completely remove the hazards to the internal validity of the test, but true experiments are more powerful, vis-à-vis other experimental designs, to reduce them (see also Bachman & Schutt, 2013, Chapter 7).

To emphasise, threats to internal validity is only one family of many types of concerns; external validity is also an issue to deal with, and randomisation alone cannot resolve issues with the generalisation of the test. Even the most perfectly executed trial can still suffer from non-generalisation issues if the contextual, historical and environmental settings have changed since the time it was executed. We used to think that the time it takes the police to arrive to the scene of a crime does not make a difference in the likelihood of apprehension of offenders and prevention of additional harm to victims, and therefore we advised against the rapid deployment of police cars (Weisburd & Eck, 2004). However, that conclusion was reached before the age of mobile telephones and more advanced deployment technologies. New evidence has emerged to demonstrate the opposite conclusion, with faster response time playing an important role in reducing injuries related to domestic violence (DeAngelo et al., 2020; Vidal & Kirchmaier, 2018).

When can we tell that the findings from one test are transferable to other settings, people or time? These are difficult questions, even though we would like to think of our theories in sociology, economy or criminology as timeless. If there is one thing we have learnt from the reproducibility project in psychology (J.R. Stevens, 2017), for example, it is that our understanding of how the world operates is fallible; theories tend to *not* be as freely transmittable as we assumed, and tests struggle to replicate and reproduce findings from original and classic experiments. Original theories do not always stand the test of time, and therefore replications are always needed to debunk criticism about the generalisability of research findings. Only through replications, diversification and localised knowledge that produce repeated findings can we make research findings sufficiently convincing (see applications in Bruinsma & Weisburd, 2007).

We then moved on to review some of the most common experimental blueprints in science. We have covered 13 common designs in Chapter 4, but there are many others. These experimental designs, ranging on a scale of their ability to remove some of the threats to the validity of the causal conclusions, all share the same purpose: to estimate the causal effect of an independent variable on dependent variables. The degree of success depends on their ability to avoid the perils discussed in Chapter 3. The choice of design ought not just to reflect conventions in research methods (e.g., Jupp, 2006) but ought also to fit the field and operational contexts in which the study is conducted – hence the reason why we do not see only RCTs in causal research.

To illustrate how experiments can overcome these difficulties, and to introduce the reader to the world of applied experimental designs, Chapter 5 discussed some necessary steps in how experiments happen. We suggested researchers use templates, reminders, nudges and blueprints in the conduct of experiments, as there is evidence to suggest that using these quality control checks and balances is advantageous (Moher et al., 2001). For example, the experimental protocol can enhance the efficiency of running trials; more importantly, they introduce more transparency, clarity and accountability in research. Similarly, templates for reporting experimental findings, such as the CONSORT Statement, are hugely expedient as they homogenise scientific reporting, nudge the authors to include all the necessary bits of information for critically assessing experiments and simplify the dissemination of research. During the running of the experiment, quality checks using field managers, especially in the form of pracademics, and knowledge gained through implementation sciences, are pivotal.

Not all experimental designs are created equal; choose the most appropriate wisely

Experimental designs are required to make valid causal estimates. A controlled test, with convincing counterfactual conditions to which the intervention is compared, is needed if we want to make reasonably sound causal expressions (Farrington, 1983). The experimental approach has the power to control the course of the trial and to single out the consequences of the manipulation of interest. Observations alone are not enough. Surveys and non-experimental designs have the power to present correlational, but not causal, relationships about the world. To estimate causal inferences, to predict consequences and to make sensible statements about how people may be affected by a particular intervention, causal designs are needed (Ariel, 2018).

Yes, there is a degree of hierarchy in causal research methods, ranging from quasi-experimental, through pre-experimental, to true experiments. Within true experiments, some models are more dependable than others. All experimenters search for as much control against possible threats to the validity and reliability of their tests – and there are plenty – but they would tend to agree that true experiments exert more control than non-true experiments. It is not easy to rule out alternative hypotheses to an observed effect, even though it is fundamental for the causal stipulation. As some methodological approaches have a better chance of removing certain threats, they should be preferred. Experiments with a prospective assignment of units into treatment and control conditions using random allocation are the answer. Relying on theorems in mathematics and probability theory, randomisation has the best chance to control many of these threats. Science has yet to produce a more robust approach for determining causation, so pre-experimental or quasi-experimental designs are a compromise when true experiments are not possible.

The principal superiority of true experiments over non-true experiments was made long ago (Fisher, 1935; McCall, 1923): experimental rather than quasi-experimental designs are preferred for causal assessment, whenever they are possible. Experiments that use statistical controls in lieu of randomisation are doing so due to practical or ethical reasons. Selection biases and misspecification errors – that is, the inability to statistically control for confounding variables – are still a major source of concern. This does not mean we should discount studies based on statistical matching techniques, but when all things are considered, carefully executed randomised controlled designs have a stronger chance of providing valid causal estimates of the treatment effect, beyond those produced through statistical models.

To emphasise, there are many situations when prospective RCTs are impossible, unethical or impractical. If the conditions are not appropriate – for example, when there is an insufficient number of cases to randomise – then randomisation is inappropriate. The test would be 'doomed to failure', and a pre-mortem analysis should have advised against running an RCT under these conditions. Non-true experimental designs should then be considered.

Furthermore, the contribution to science from *any* experiment is incontestable. There are learning points from every study. We have repeatedly tried to make the case for viewing causal expressions as building blocks of knowledge. No stand-alone trial is enough. Yet if replications of a test produce a series of reliable outcomes, or pre-experimental designs reproduce similar findings as the literature indicated, then why should we immediately discount them? There are different ways to isolate the causal factor from alternative predictors, and when considered holistically, evidence from all experimental designs is important, even if not all evidence is created equal (Ariel, 2019).

Sometimes, however, we have a limited body of evidence on a particular research question, and we are then forced to critically assess the level of evidence produced against a benchmark. The production of the evidence will be evaluated against the model used – and how well – as a stand-alone piece of research, the test convinces us of its conclusion. How well would a quasi-experiment control for all the threats to the validity of the test? Can a pretest–post-test-only experiment reduce our anxiety about selection biases? Would one experiment with only two units (e.g. one treatment vs one control hotspot) produce valid estimates that are sufficiently strong to dictate future policing policy? Categorically, the answers to these questions are no. The risk is too great.

The future of experimental disciplines

Despite the hierarchal relationship we just portrayed about the different experimental designs within causal research, how true experiments are viewed by different social scientists is an ongoing debate. If RCTs are the 'gold standard',[1] why do we not see more of them in applied sciences, such as criminology, education or political sciences? After all, there are just a few hundred randomised experiments in policing, and even fewer in the political sciences. As the theory behind experimental models is not contested in mainstream quantitative research (though their applicability is), why would they not be more prevalent in the social sciences?

In the remaining pages, we try to unearth some of the reasons why we do not have a blossoming community of experimenters in these sciences, as we see, for instance, in the biomedical disciplines. We contextualise these reasons within three trends: proliferation, diversification and collaboration. These discussions will lead us to suggestions on how to mainstream experiments into the social sciences more robustly.

Proliferation

As the experimental community grows and more social science researchers apply experimental methods to the study of crime and crime policy, the sheer number of randomised trials is rapidly accumulating on a numerical basis. Rigorous evaluations are increasingly required, particularly during periods of public expenditure austerity. We simply cannot afford to do the same 'things' that may not provide any benefit with the current budget cuts to crime policies. This is a global phenomenon. Yet, these financial crises should not be wasted and can be viewed as an opportunity: our experience with policymakers around the globe is that they all recognise that practices must be subject to rigorous testing. Expensive policies and tactics that only a decade ago would be implemented based on hunches and goodwill are now required to undergo piloting, impact evaluations and academic assessment prior to full deployment.

At this juncture, experimental disciplines shine. As more police forces, justice departments, judiciaries and treatment providers undergo these similar processes, more collaborations between these agencies and experimenters emerge, the number of research projects increases and we move closer to a vision of experimental disciplines in which major policy decisions are not taken in the absence of multiple experiments on an issue (Sherman & Cohn, 1989).

---

[1]The popular title 'gold standard' may be unfitting (see arguments by Sampson, 2010), although we still argue that, in principle, the hierarchical nature of research methods remains intact.

Diversification

With the proliferation and infiltration of experimental work into decision-making comes a wide gamut of research questions. Evident from George Mason's excellent *Centre for Evidence-Based Crime Policy*, we see an expansion of the units of analysis, population types and treatment categories under investigation. This is true for all research methods, but it seems particularly true for experimental designs. For instance, proactive, rather than reactive, crime policies – embodied chiefly in mature police departments such as the West Midlands Police and Philadelphia Police Department – seem to introduce a new model for policymaking: *test first, apply later*. New methods of preventing crime, managing criminals, caring for vulnerable populations and focusing on hotspots are constantly mushrooming, and within this general movement, it is no wonder that experimental criminology is rapidly increasing worldwide.

Randomised trials provide valid causal estimates of treatment effects, which is why there is a strong consensus that experiments are the 'gold standard' for evaluation research in criminology. That which works, does not work or is (currently) only promising given limited evidence can be effectively demonstrated through proper tests. The number of research questions, mirroring the various tactics that can be effectively analysed experimentally, are put to the test prior to procurement or force-wide implementation. Body-worn cameras, recruiters of criminal networks, honeypots, soft policing in hotspots, parental patrols in hotspots, GPS and WiFi tracking of officers, legitimacy in counterterrorism and so on mirror this diversification (Sherman et al., 2014; Wain et al., 2017).

Collaboration

However, the state of the art of experimental fields can improve in two major areas, which we feel have transferrable implications for other fields developing experimental approaches. As evident in Braga et al.'s (2014) work on experimenters' networks, 'neighbourhoods' are small and disconnected. Compared perhaps to experimental psychology, the size of co-authorship and the number of cross-institutional collaborations should equally grow. The quantity of ties each node – or experimenter – is associated with is often limited, except those neighbourhoods associated with Professor Lawrence Sherman and Professor David Weisburd. We believe that with new experimenters covering new grounds, indeed, the network will expand with a greater number of degrees of connection between institutions and researchers. Thus, the more we talk to one another, the better. Yet to do this, we need more tutoring and a greater willingness by supervisors to work with junior experimenters on their trials. The Jerry Lee Centre for Experimental Criminology (https://www.crim.cam.ac.uk/research/research-centres/experimental-criminology)

and Cambridge University's Master of Studies in Applied Criminology and Police Management (https://www.crim.cam.ac.uk/Courses/mst-courses/MStPolice) programme are exactly the kinds of models we need in order to expand more globally.

How do we mainstream experiments into policymaking processes?

Linked to our need for more influential nodes in experimental criminology, a major issue that we face is the misalignments between the moving parts associated with controlled trials. As we all know, experiments require great(er) attention to proper planning and design before the first participant is rolled into the study. In field trials, where the stakes are high, this process can take months, possibly years, of piloting and recalibrating, until the proper protocol is established. Yet, this process does not always correspond with institutional processes, procurement cycles or how long the decision-maker has before taking on a new role within an organisation. As important, funding cycles nearly always do not match the needs of the experiment, which in many ways leaves the game to senior experimenters who can divert available research funds into new projects. If a chief of police asked us to conduct an experiment on body-worn cameras and we had to wait for a funding agency to provide resources in 12 to 18 months, clearly the opportunity to conduct the experiment would vanish.

Finally, a stronger link between research institutions and in-house pracademics is required. As we discussed in Chapter 5, it will not only widen the network of interested partnerships, and cement our role in evidence-based policy, but it will also help to streamline experiments. Our experience with the British Transport Police (BTP) and the leadership of (then) Assistant Chief Constable Mark Newton is an example of an incredible academic–practitioner relationship. BTP has materialised Sherman's model of 'Totally Evidence-Based Policing Agency' (Sherman, 2015) with more experiments under way, a strong steering towards using the scientific approach in policing and continuous learning. This is the 'scientification' of policing. A strong and dedicated team of analysts constantly seek 'testable' questions in a policing environment and then design, conduct and analyse the results prior to deployment. Only time will tell whether the infiltration of pracademics into experimental disciplines will be a game changer in the (much-needed) proliferation of experimental designs in policymaking.

Further Reading

Farrington, D. P., Lösel, F., Boruch, R. F., Gottfredson, D. C., Mazerolle, L., Sherman, L. W., & Weisburd, D. (2019). Advancing knowledge about replication in criminology. *Journal of Experimental Criminology*, *15*(3), 373–396.

This chapter discussed issues associated with the generalisability of research findings, especially given the growing body of evidence that is *unable* to replicate findings in many experiments in psychology. These concerns have been raised in a special edited volume in the *Journal of Experimental Criminology*, with a particular focus on criminology. This article by Farrington et al. summarises the issues raised in the volume and should be consulted.

GLOSSARY

Bias: An intentional or unintentional systematic error in an estimate. Bias leads to inaccuracy in measurement and usually results in misleading inferences about the relationship between the variables.

Causal inference: Conclusions drawn from a sample or specific study conditions about the causal relationship between two or more phenomena. For valid causal inference, three conditions must be met: (1) a statistical correlation between the two variables, (2) a temporal sequence such that the cause precedes the effect and (3) the absence of an alternative explanation other than the independent variable for the observed change in the dependent variable.

Causality: A relationship between two or more variables whereby change in one variable results in a reaction in the other variables(s).

Control group: A comparison group that is unexposed to the studied treatment effect. In experiments, participants can be assigned to the control group either randomly (by chance) or using statistical matching techniques when randomisation is not possible. Control groups can comprise of no-treatment, placebo or alternative treatments.

Counterfactual: A 'parallel universe' with identical conditions to the treatment group but without the treatment applied.

Covariate: A variable associated with the outcome variable that can therefore affect the relationship between the studied intervention and the outcome variable. These extraneous variables are included in quasi-experimental designs to rule out alternative explanations to the observed change in the outcome variable, as well as to increase the precision of the overall causal model.

Dependent variable: The variable affected by the independent variable; the outcome of the stimulus applied in an experiment.

Effect size: The magnitude of the difference between treatment and control conditions following the intervention, expressed in standardised units.

Effectiveness: An expression of the benefit of an intervention measured under 'real-world' but controlled experimental settings.

Efficacy: An expression of the benefit of an intervention, measured under ideal and controlled experimental settings.

Evidence-based policy: The use of scientific method to produce policy recommendations.

External validity: The degree to which the study outcomes can be generalised to different people, places, times and contexts; often expressed in narrative rather than mathematical terms.

Implementation: A set of processes that have taken place (or that have been withheld) as an indispensable part of the studied treatment and its effects.

Independent variable: The variable that causes a change in the dependent variable; the stimulus or treatment applied in an experiment.

Intention to treat: An analytical approach in randomised controlled trials where the units are assumed to have been exposed to the condition to which they were assigned; ignores any violations of the allocation sequence or the level of completion of the assigned treatment.

Interaction effect: Situations where two or more variables jointly affect the dependent variable, thus considered a new treatment term.

Internal validity: The degree to which the inference about the causal relationship between the independent and dependent variables is valid.

Matching: A statistical approach in which a balanced comparison group is created based on the pretreatment characteristics of the treatment group participants; unlike randomisation, in which balance is gained through the random allocation of participants into treatment and control conditions. Statistical matching can be performed on measured data but cannot control for differences between treatment and control conditions based on unmeasured data.

Natural experiment: A methodological approach to identify causal inference in which cause-and-effect relationships are observed in their natural settings, without the direct involvement of the researcher in the form of allocating units into treatment and control conditions or applying the intervention.

Null hypothesis: A statement about the lack of a relationship between the variables under investigation. As the causal expression that is tested in the experiment, the null hypothesis serves as the starting point in experimental research.

Observational research: A research design in which phenomena are described without drawing inferences about causal relationships. Observational studies involve only measurement, not manipulation of stimuli.

Participant: Any type of unit that takes part in a study, such as individuals, cases or groups.

Publication bias: The systematic error associated with selective dissemination of results, when findings that reject the null hypothesis are more likely to be published. Also known as the 'file-drawer' problem, publication bias often can lead to errors in systematic reviews and meta-analyses because the overall results may erroneously suggest that the treatment is more effective than it is.

Quasi-experiment: A causal design in which participants are not randomly assigned to treatment and control conditions.

Random sampling: The procedure of selecting a group of participants out of the population using chance alone, giving every participant in the studied population the same probability of being recruited into the study sample.

Randomisation: The allocation of units into treatment and control conditions based on chance. Over time and with sufficiently large samples, random allocation creates balanced treatment and control groups before administrating the treatment in one but not in the other group (i.e. at baseline).

Selection bias: A process of selecting units to participate in an experiment that results in unbalanced groups, usually when the allocation systematically favours one group over the other; often leads to invalid inferences about the treatment effect.

Specification error: Bias resulting from an incomplete or imprecise statistical model of causality, usually due to omitted control variables or imprecise measurements.

Statistical power: The likelihood that an experiment will be able to appropriately reject the null hypothesis – that is, the ability of an experiment to detect a statistically significant effect that exists in the population.

Stimulus (plural stimuli): The intervention or treatment that the experimenter manipulates (or observed in natural experimental settings).

Systematic review: A type of literature review that aims to collect information on all published and unpublished studies relating to a particular research question. Unlike narrative reviews that are often subjective, systematic reviews are objective, with greater transparency and systematic methodology about how evidence was collected and synthesised.

Time-series analysis: A statistical approach in which multiple waves of observations of the data are made chronologically. In causal research, time-series analyses often explore how a trend in the dependent variable was 'interrupted' by a treatment effect (also known as interrupted time-series analysis).

Trickle flow assignment: A process of randomly allocating eligible units into treatment and control conditions over time, as the units become available – as opposed to batch random assignment, in which all units are randomly assigned simultaneously.

Check out the next title in the collection: *Linear Regression: An Introduction to Statistical Models*, **for guidance on Linear Regression, the fundamental statistical model used in quantitative social research.**

REFERENCES

Abadie, A., & Gardeazabal, J. (2003). The economic costs of conflict: A case study of the Basque Country. *American Economic Review, 93*(1), 113–132. https://doi.org/10.1257/000282803321455188

Abell, P., & Engel, O. (2019). Subjective causality and counterfactuals in the social sciences: Toward an ethnographic causality? *Sociological Methods & Research.* Advance online publication. https://doi.org/10.1177/0049124119852373

Abend, G., Petre, C., & Sauder, M. (2013). Styles of causal thought: An empirical investigation. *American Journal of Sociology, 119*(3), 602–654. https://doi.org/10.1086/675892

Abou-El-Fotouh, H. A. (1976). Relative efficiency of the randomised complete block design. *Experimental Agriculture, 12*(2), 145–149. https://doi.org/10.1017/S0014479700007213

Abramowitz, M., & Stegun, I. A. (Eds.). (1972). *Handbook of mathematical functions with formulas, graphs, and mathematical tables* (Vol. 55, 10th Printing). National Bureau of Standards.

Alderman, T. (2020). Can a police-delivered intervention enhance students' online safety? A cluster randomised controlled trial on the effect of the ThinkUKnow programme in the Australian Capital Territory [Unpublished master's dissertation]. University of Cambridge.

Allen, M. (Ed.). (2017). *The SAGE encyclopaedia of communication research methods.* Sage. https://doi.org/10.4135/9781483381411

Alm, J. (1991). A perspective on the experimental analysis of taxpayer reporting. *Accounting Review, 66*(3), 577–593.

Altman, D. G. (1985). Comparability of randomised groups. *Journal of the Royal Statistical Society: Series D (The Statistician), 34*(1), 125–136. https://doi.org/10.2307/2987510

Altman, D. G. (1990). *Practical statistics for medical research.* CRC Press. https://doi.org/10.1201/9780429258589

Altman, D. G. (1991). Statistics in medical journals: Developments in the 1980s. *Statistics in Medicine, 10*(12), 1897–1913. https://doi.org/10.1002/sim.4780101206

Altman, D. G. (1996). Better reporting of randomised controlled trials: The CONSORT statement. *British Medical Journal, 313*(7057), 570–571. https://doi.org/10.1136/bmj.313.7057.570

Altman, D. G., & Doré, C. J. (1990). Randomisation and baseline comparisons in clinical trials. *The Lancet, 335*(8682), 149–153. https://doi.org/10.1016/0140-6736(90)90014-V

Amendola, K. L., Weisburd, D., Hamilton, E. E., Jones, G., & Slipka, M. (2011). An experimental study of compressed work schedules in policing: Advantages and disadvantages of various shift lengths. *Journal of Experimental Criminology, 7*(4), 407–442. https://doi.org/10.1007/s11292-011-9135-7

Amendola, K. L., & Wixted, J. T. (2015). Comparing the diagnostic accuracy of suspect identifications made by actual eyewitnesses from simultaneous and sequential lineups in a randomized field trial. *Journal of Experimental Criminology, 11*(2), 263–284. https://doi.org/10.1007/s11292-014-9219-2

American Psychological Association, American Educational Research Association, & National Council on Measurement in Education. (1954). Technical recommendations for psychological tests and diagnostic techniques (Vol. *51*, No. 2). American Psychological Association.

Angel, C. M., Sherman, L. W., Strang, H., Ariel, B., Bennett, S., Inkpen, N., Keane, A., & Richmond, T. S. (2014). Short-term effects of restorative justice conferences on post-traumatic stress symptoms among robbery and burglary victims: A randomised controlled trial. *Journal of Experimental Criminology, 10*(3), 291–307. https://doi.org/10.1007/s11292-014-9200-0

Angelucci, M., & Di Maro, V. (2016). Programme evaluation and spillover effects. *Journal of Development Effectiveness, 8*(1), 22–24. https://doi.org/10.1080/19439342.2015.1033441

Angrist, J. D. (2006). Instrumental variables methods in experimental criminological research: What, why and how. *Journal of Experimental Criminology, 2*(1), 23–44. https://doi.org/10.1007/s11292-005-5126-x

Angrist, J. D., & Pischke, J. S. (2014). *Mastering 'metrics: The path from cause to effect.* Princeton University Press.

Antrobus, E., Elffers, H., White, G., & Mazerolle, L. (2013). Nonresponse bias in randomised controlled experiments in criminology: Putting the Queensland Community Engagement Trial (QCET) under a microscope. *Evaluation Review, 37*(3–4), 197–212. https://doi.org/10.1177/0193841X13518534

Antrobus, E., Thompson, I., & Ariel, B. (2019). Procedural justice training for police recruits: Results of a randomised controlled trial. *Journal of Experimental Criminology, 15*(1), 29–53. https://doi.org/10.1007/s11292-018-9331-9

Apel, R. J., & Sweeten, G. (2010). Propensity score matching in criminology and criminal justice. In A. Piquero & D. Weisburd (Eds.), *Handbook of quantitative criminology* (pp. 543–562). Springer. https://doi.org/10.1007/978-0-387-77650-7_26

Apospori, E., & Alpert, G. (1993). Research note: The role of differential experience with the criminal justice system in changes in perceptions of severity of legal sanctions over time. *Crime & Delinquency, 39*(2), 184–194. https://doi.org/10.1177/0011128793039002004

Ariel, B. (2009, January). Systematic review of baseline imbalances in randomised controlled trials in criminology [Paper presentation]. The communicating complex statistical evidence conference, University of Cambridge, UK.

Ariel, B. (2011, July 5). London underground crime data (2009–2011) & Operation "BTP-LU-RCT" [Paper presentation]. The 4th international NPIA-Cambridge conference on evidence-based policing, Cambridge, UK.

Ariel, B. (2012). Deterrence and moral persuasion effects on corporate tax compliance: Findings from a randomised controlled trial. *Criminology, 50*(1), 27–69. https://doi.org/10.1111/j.1745-9125.2011.00256.x

Ariel, B. (2016). Police body cameras in large police departments. *Journal of Criminal Law & Criminology, 106*(4), 729–768.

Ariel, B. (2018). Not all evidence is created equal: On the importance of matching research questions with research methods in evidence based policing. In R. Mitchell & L. Huey (Eds.), *Evidence based policing: An introduction* (pp. 63–86). Policy Press.

Ariel, B. (2019). Technology in policing. In D. Weisburd & A. A. Braga (Eds.), *Innovations in policing: Contrasting perspectives* (2nd ed., pp. 521–516). Cambridge University Press.

Ariel, B., & Bland, M. (2019). Is crime rising or falling? A comparison of police-recorded crime and victimization surveys. *Methods of Criminology and Criminal Justice Research (Sociology of Crime, Law and Deviance, Vol. 24,* pp. 7-31). Emerald Publishing Limited.

Ariel, B., Bland, M., & Sutherland, A. (2017). 'Lowering the threshold of effective deterrence'—Testing the effect of private security agents in public spaces on crime: A randomized controlled trial in a mass transit system. *PLOS ONE, 12*(12), e0187392.

Ariel, B., Englefield, A., & Denley, J. (2019). "I heard it through the grapevine": A randomised controlled trial on the direct and vicarious effects of preventative specific deterrence initiatives in criminal networks. *Journal of Criminal Law & Criminology, 109*(4), 819–867.

Ariel, B., & Farrar, W. (2012). The Rialto Police Department wearable cameras experiment experimental protocol: CRIMPORT. Institute of Criminology, University of Cambridge.

Ariel, B., Farrar, W. A., & Sutherland, A. (2015). The effect of police body-worn cameras on use of force and citizens' complaints against the police: A randomised controlled trial. *Journal of Quantitative Criminology*, *31*(3), 509–535. https://doi.org/10.1007/s10940-014-9236-3

Ariel, B., & Farrington, D. P. (2014). Randomised block designs. In G. Bruinsma & D. Weisburd (Eds.), *Encyclopedia of criminology and criminal justice* (pp. 4273–4283). Springer. https://doi.org/10.1007/978-1-4614-5690-2_52

Ariel, B., Garner, G., Strang, H., & Sherman, L. W. (2019, June 6). Creating a critical mass for a global movement in evidence-based policing: The Cambridge Pracademia [Paper Presentation]. 2019 Drapkin Symposium, Hebrew University, Jerusalem, Israel.

Ariel, B., & Langley, B. (2019, July 10). Procedural justice in preventing terrorism: An RCT and rollout [Paper presentation]. The 12th international evidence-based policing conference, Cambridge, UK.

Ariel, B., Lawes, D., Weinborn, C., Henry, R., Chen, K., & Brants Sabo, H. (2019). The 'less-than-lethal weapons effect' – Introducing TASERs to routine police operations in England and Wales: A randomised controlled trial. *Criminal Justice and Behavior*, *46*(2), 280–300. https://doi.org/10.1177/0093854818812918

Ariel, B., Mitchell, R. J., Tankebe, J., Firpo, M. E., Fraiman, R., & Hyatt, J. M. (2020). Using wearable technology to increase police legitimacy in Uruguay: The case of body-worn cameras. *Law & Social Inquiry*, *45*(1), 52–80. https://doi.org/10.1017/lsi.2019.13

Ariel, B., Newton, M., McEwan, L., Ashbridge, G. A., Weinborn, C., & Brants, H. S. (2019). Reducing assaults against staff using body-worn cameras (BWCs) in railway stations. *Criminal Justice Review*, *44*(1), 76–93. https://doi.org/10.1177/0734016818814889

Ariel, B., & Partridge, H. (2017). Predictable policing: Measuring the crime control benefits of hotspots policing at bus stops. *Journal of Quantitative Criminology*, *33*(4), 809–833. https://doi.org/10.1007/s10940-016-9312-y

Ariel, B., & Sherman, L. W. (2012). Mandatory arrest for misdemeanour domestic violence effects on repeat offending: Protocol (1–30). *Campbell Systematic Reviews*, *8*(1), 1–30. https://doi.org/10.1002/CL2.85

Ariel, B., Sherman, L. W., & Newton, M. (2020). Testing hot-spots police patrols against no-treatment controls: Temporal and spatial deterrence effects in the London Underground experiment. *Criminology*, *58*(1), 101–128. https://doi.org/10.1111/1745-9125.12231

Ariel, B., Sutherland, A., & Bland, M. (2019). The trick does not work if you have already seen the gorilla: How anticipatory effects contaminate pre-treatment measures in field experiments. *Journal of Experimental Criminology.* Advance online publication. https://doi.org/10.1007/s11292-019-09399-6

Ariel, B., Sutherland, A., Henstock, D., Young, J., Drover, P., Sykes, J., Megicks, S., & Henderson, R. (2016a). Report: Increases in police use of force in the presence of body-worn cameras are driven by officer discretion: A protocol-based subgroup analysis of ten randomised experiments. *Journal of Experimental Criminology, 12*(3), 453–463. https://doi.org/10.1007/s11292-016-9261-3

Ariel, B., Sutherland, A., Henstock, D., Young, J., Drover, P., Sykes, J., Megicks, S., & Henderson, R. (2016b). Wearing body cameras increases assaults against officers and does not reduce police use of force: Results from a global multi-site experiment. *European Journal of Criminology, 13*(6), 744–755. https://doi.org/10.1177/1477370816643734

Ariel, B., Sutherland, A., Henstock, D., Young, J., Drover, P., Sykes, J., Megicks, S., & Henderson, R. (2017). "Contagious accountability": A global multisite randomised controlled trial on the effect of police body-worn cameras on citizens' complaints against the police. *Criminal Justice and Behavior, 44*(2), 293–316. https://doi.org/10.1177/0093854816668218

Ariel, B., Sutherland, A., & Sherman, L. W. (2019). Preventing treatment spillover contamination in criminological field experiments: The case of body-worn police cameras. *Journal of Experimental Criminology, 15*(4), 569–591. https://doi.org/10.1007/s11292-018-9344-4

Ariel, B., Vila, J., & Sherman, L. (2012). Random assignment without tears: how to stop worrying and love the Cambridge randomizer. *Journal of Experimental Criminology, 8*(2), 193–208.

Ariel, B., Weinborn, C., & Sherman, L. W. (2016). "Soft" policing at hot spots: Do police community support officers work? A randomised controlled trial. *Journal of Experimental Criminology, 12*(3), 277–317. https://doi.org/10.1007/s11292-016-9260-4

Assmann, S., Pocock, S., Enos, L., & Kasten, L. (2000). Subgroup analysis and other (mis)uses of baseline data in clinical trials. *The Lancet, 355*(9209), 1064–1069. https://doi.org/10.1016/S0140-6736(00)02039-0

Babcock, J. C., Green, C. E., & Robie, C. (2004). Does batterers' treatment work? A meta-analytic review of domestic violence treatment. *Clinical Psychology Review, 23*(8), 1023–1053. https://doi.org/10.1016/j.cpr.2002.07.001

Bacchieri, A., & Della Cioppa, G. (2007). *Fundamentals of clinical research: Bridging medicine, statistics and operations.* Springer Science & Business Media. https://doi.org/10.1007/978-88-470-0492-4

Bachman, R., & Schutt, R. K. (2013). *The practice of research in criminology and criminal justice.* Sage.

Baird, S., Bohren, J. A., McIntosh, C., & Özler, B. (2014). *Designing experiments to measure spillover effects.* The World Bank. https://doi.org/10.1596/1813-9450-6824

Baker, M. (2016). Reproducibility crisis. *Nature, 533*(7604), 353–366.

Barnard, J., Du, J., Hill, J. L., & Rubin, D. B. (1998). A broader template for analyzing broken randomised experiments. *Sociological Methods & Research, 27*(2), 285–317. https://doi.org/10.1177/0049124198027002005

Barnes, G. C., Ahlman, L., Gill, C., Sherman, L. W., Kurtz, E., & Malvestuto, R. (2010). Low-intensity community supervision for low-risk offenders: A randomised, controlled trial. *Journal of Experimental Criminology, 6*(2), 159–189. https://doi.org/10.1007/s11292-010-9094-4

Barnes, G. C., Hyatt, J. M., & Sherman, L. W. (2017). Even a little bit helps: An implementation and experimental evaluation of cognitive-behavioral therapy for high-risk probationers. *Criminal Justice and Behavior, 44*(4), 611–630. https://doi.org/10.1177/0093854816673862

Barnes, G. C., Williams, S., Sherman, L. W., Parmar, J., House, P., & Brown, S. A. (2020). *Sweet spots of residual deterrence: A randomized crossover experiment in minimalist police patrol.* SocArXiv. https://doi.org/10.31235/osf.io/kwf98

Bartos, B. J., McCleary, R., Mazerolle, L., & Luengen, K. (2020). Controlling gun Violence: Assessing the impact of Australia's Gun Buyback Program using a synthetic control group experiment. *Prevention Science, 21*(1), 131–136.

Baumeister, R. F., & Leary, M. R. (1997). Writing narrative literature reviews. *Review of General Psychology, 1*(3), 311–320. https://doi.org/10.1037/1089-2680.1.3.311

Beebee, H., Hitchcock, C., & Menzies, P. (Eds.). (2009). *The Oxford handbook of causation.* Oxford University Press. https://doi.org/10.1093/oxfor dhb/9780199279739.001.0001

Beller, E. M., Gebski, V., & Keech, A. C. (2002). Randomisation in clinical trials. *Medical Journal of Australia, 177*(10), 565–567. https://doi. org/10.5694/j.1326-5377.2002.tb04955.x

Bellg, A. J., Borrelli, B., Resnick, B., Hecht, J., Minicucci, D. S., Ory, M., Ogedegbe, G., Orwig, D., Ernst, D., & Czajkowski, S. (2004). Enhancing treatment fidelity in health behavior change studies: Best practices and recommendations from the NIH Behavior Change Consortium. *Health Psychology, 23*(5), 443–451. https://doi. org/10.1037/0278-6133.23.5.443

Bennett, S., Mazerolle, L., Antrobus, E., Eggins, E., & Piquero, A. R. (2018). Truancy intervention reduces crime: Results from a randomised field trial. *Justice Quarterly, 35*(2), 309–329. https://doi.org/10.1080/07418825.2017.1313440

Bennett, S., Newman, M., & Sydes, M. (2017). Mobile police community office: A vehicle for reducing crime, crime harm and enhancing police legitimacy? *Journal of Experimental Criminology, 13*(3), 417–428. https://doi.org/10.1007/s11292-017-9302-6

Berger, V. W. (2004). Selection bias and baseline imbalances in randomised trials. *Drug Information Journal, 38*(1), 1–2. https://doi.org/10.1177/009286150403800101

Berger, V. W. (2005a). Is allocation concealment a binary phenomenon? *Medical Journal of Australia, 183*(3), 165–166. https://doi.org/10.5694/j.1326-5377.2005.tb06974.x

Berger, V. W. (2005b). *Selection bias and covariate imbalances in randomised clinical trials.* Wiley. https://doi.org/10.1002/0470863641

Berger, V. W. (2006). Varying the block size does not conceal the allocation. *Journal of Critical Care, 21*(2), 299. https://doi.org/10.1016/j.jcrc.2006.01.002

Berger, V. W., & Exner, D. V. (1999). Detecting selection bias in randomised clinical trials. *Controlled Clinical Trials, 20*(4), 319–327. https://doi.org/10.1016/S0197-2456(99)00014-8

Berger, V. W., & Weinstein, S. (2004). Ensuring the comparability of comparison groups: Is randomisation enough? *Control Clinical Trials, 25*(5), 515–524. https://doi.org/10.1016/j.cct.2004.04.001

Berk, R. A. (1988). Causal inference for sociological data. In N. J. Smelser (Ed.), *Handbook of sociology* (pp. 155–172). Sage.

Berk, R. A., & Sherman, L. W. (1985). Data collection strategies in the Minneapolis domestic assault experiment. In L. Burstein, H. E. Freeman, & P. H. Rossi (Eds.), *Collecting evaluation data: Problems and solutions* (pp. 35–48). Sage.

Berk, R. A., Sorenson, S. B., & Barnes, G. (2016). Forecasting domestic violence: A machine learning approach to help inform arraignment decisions. *Journal of Empirical Legal Studies, 13*(1), 94–115. https://doi.org/10.1111/jels.12098

Bernal, J. L., Cummins, S., & Gasparrini, A. (2017). Interrupted time series regression for the evaluation of public health interventions: A tutorial. *International Journal of Epidemiology, 46*(1), 348–355.

Bernard, H. R. (2017). *Research methods in anthropology: Qualitative and quantitative approaches.* Rowman & Littlefield.

Bilach, T. J., Roche, S. P., & Wawro, G. J. (2020). The effects of the summer All Out Foot Patrol Initiative in New York City: A difference-in-differences approach. *Journal of Experimental Criminology.* Advance online publication. https://doi.org/10.1007/s11292-020-09445-8

Bilderbeck, A. C., Farias, M., Brazil, I. A., Jakobowitz, S., & Wikholm, C. (2013). Participation in a 10-week course of yoga improves behavioural control and decreases psychological distress in a prison population. *Journal of Psychiatric*

Research, 47(10), 1438–1445. https://doi.org/10.1016/j.jpsychires.2013.06.014

Blackwell, D., & Hodges, J. L. (1957). Design for the control of selection bias. *Annals of Mathematical Statistics, 28*(2), 449–460. https://doi.org/10.1214/aoms/1177706973

Bland, M. P., & Ariel, B. (2015). Targeting escalation in reported domestic abuse: Evidence from 36,000 callouts. *International Criminal Justice Review, 25*(1), 30–53. https://doi.org/10.1177/1057567715574382

Bland, M. P., & Ariel, B. (2020). *Targeting domestic abuse with police data.* Springer Nature. https://doi.org/10.1007/978-3-030-54843-8

Bolzern, J. E., Mitchell, A., & Torgerson, D. J. (2019). Baseline testing in cluster randomised controlled trials: Should this be done? *BMC Medical Research Methodology, 19*(1), Article 106. https://doi.org/10.1186/s12874-019-0750-8

Bottoms, A., & Tonry, M. (Eds.). (2013). *Ideology, crime and criminal justice.* Routledge.

Boruch, R. F. (1997). *Randomised experiments for planning and evaluation.* Sage. https://doi.org/10.4135/9781412985574

Boruch, R. F., Snyder, B., & DeMoya, D. (2000). The importance of randomised field trials. *Crime & Delinquency, 46*(2), 156–180. https://doi.org/10.1177/0011128700046002002

Boruch, R. F., Victor, T., & Cecil, J. S. (2000). Resolving ethical and legal problems in randomised experiments. *Crime & Delinquency, 46*(3), 330–353. https://doi.org/10.1177/0011128700046003005

Bowers, J., Desmarais, B. A., Frederickson, M., Ichino, N., Lee, H. W., & Wang, S. (2018). Models, methods and network topology: Experimental design for the study of interference. *Social Networks, 54*, 196–208. https://doi.org/10.1016/j.socnet.2018.01.010

Braga, A. A. (2016). The value of "pracademics" in enhancing crime analysis in police departments. *Policing, 10*(3), 308–314. https://doi.org/10.1093/police/paw032

Braga, A. A., & Bond, B. J. (2008). Policing crime and disorder hot spots: A randomised controlled trial. *Criminology, 46*(3), 577–607. https://doi.org/10.1111/j.1745-9125.2008.00124.x

Braga, A. A., & Hinkle, M. (2010). The participation of academics in the criminal justice working group process. In J. Klofas, N. Hipple, & E. McGarrell (Eds.), *The new criminal justice* (pp. 114–120). Routledge.

Braga, A. A., Kennedy, D. M., Waring, E. J., & Piehl, A. M. (2001). Problem-oriented policing, deterrence, and youth violence: An evaluation of Boston's Operation Ceasefire. *Journal of Research in Crime and Delinquency, 38*(3), 195–225. https://doi.org/10.1177/0022427801038003001

Braga, A. A., Papachristos, A., & Hureau, D. (2012). Hot spots policing effects on crime. *Campbell Systematic Reviews, 8*(1), 1–96. https://doi.org/10.4073/csr.2012.8

Braga, A. A., Pierce, G. L., McDevitt, J., Bond, B. J., & Cronin, S. (2008). The strategic prevention of gun violence among gang-involved offenders. *Justice Quarterly, 25*(1), 132–162. https://doi.org/10.1080/07418820801954613

Braga, A. A., Turchan, B. S., Papachristos, A. V., & Hureau, D. M. (2019). Hot spots policing and crime reduction: An update of an ongoing systematic review and meta-analysis. *Journal of Experimental Criminology, 15*(3), 289–311. https://doi.org/10.1007/s11292-019-09372-3

Braga, A. A., & Weisburd, D. L. (2010). *Policing problem places: Crime hot spots and effective prevention.* Oxford University Press. https://doi.org/10.1093/acprof:oso/9780195341966.001.0001

Braga, A. A., Weisburd, D. L., & Turchan, B. (2019). Focused deterrence strategies effects on crime: A systematic review. *Campbell Systematic Reviews, 15*(3), Article e1051. https://doi.org/10.1002/cl2.1051

Braga, A. A., Weisburd, D. L., Waring, E. J., Mazerolle, L. G., Spelman, W., & Gajewski, F. (1999). Problem-oriented policing in violent crime places: A randomised controlled experiment. *Criminology, 37*(3), 541–580. https://doi.org/10.1111/j.1745-9125.1999.tb00496.x

Braga, A. A., Welsh, B. C., Papachristos, A. V., Schnell, C., & Grossman, L. (2014). The growth of randomised experiments in policing: The vital few and the salience of mentoring. *Journal of Experimental Criminology, 10*(1), 1–28. https://doi.org/10.1007/s11292-013-9183-2

Braithwaite, J., & Makkai, T. (1994). Trust and compliance. *Policing and Society, 4*(1), 1–12. https://doi.org/10.1080/10439463.1994.9964679

Brantingham, P. L., & Brantingham, P. J. (1999). A theoretical model of crime hot spot generation. *Studies on Crime & Crime Prevention, 8*(1), 7–26.

Brants-Sabo, H., & Ariel, B. (2020). Evidence map of school-based violence prevention programs in Israel. *International Criminal Justice Review.* Advance online publication. https://doi.org/10.1177/1057567720967074

Braucht, G. N., & Reichardt, C. S. (1993). A computerized approach to trickle-process, random assignment. *Evaluation Review, 17*(1), 79–90. https://doi.org/10.1177/0193841X9301700106

Braver, M. W., & Braver, S. L. (1988). Statistical treatment of the Solomon four-group design: A meta-analytic approach. *Psychological Bulletin, 104*(1), 150–154. https://doi.org/10.1037/0033-2909.104.1.150

Britt, C. L., & Weisburd, D. (2010). Statistical power. In A. Piquero & D. Weisburd (Eds.), *Handbook of quantitative criminology* (pp. 313–332). Springer. https://doi.org/10.1007/978-0-387-77650-7_16

Bruinsma, G. J., & Weisburd, D. (2007). Experimental and quasi-experimental criminological research in the Netherlands. *Journal of Experimental Criminology, 3*(2), 83–88. https://doi.org/10.1007/s11292-007-9032-2

Bryman, A. E. (2016). *Social research methods*. Oxford University Press.

Cameli, M., Novo, G., Tusa, M., Mandoli, G. E., Corrado, G., Benedetto, F., Antonini-Canterin, F., & Citro, R. (2018). How to write a research protocol: Tips and tricks. *Journal of Cardiovascular Echography, 28*(3), 151–153. https://doi.org/10.4103/jcecho.jcecho_41_18

Campbell, D. T. (1957). Factors relevant to the validity of experiments in social settings. *Psychological Bulletin, 54*(4), 297–312. https://doi.org/10.1037/h0040950

Campbell, D. T. (1968). The Connecticut crackdown on speeding-time-series data in quasi-experimental analysis. *Law & Society Review, 3*(1), 33–54. https://doi.org/10.2307/3052794

Campbell, D. T. (1969). Reforms as experiments. *American Psychologist, 24*(4), 409–429. https://doi.org/10.1037/h0027982

Campbell, D. T., & Boruch, R. F. (1975). Making the case for randomised assignment to treatments by considering the alternatives: Six ways in which quasi-experimental evaluations in compensatory education tend to underestimate effects. In C. A. Bennett & A. A. Lumsdaine (Eds.), *Evaluation and experiment: Some critical issues in assessing social programs* (pp. 195–296). Academic Press. https://doi.org/10.1016/B978-0-12-088850-4.50008-2

Campbell, D. T., & Kenny, D. A. (1999). *A primer on regression artifacts*. Guilford Press.

Campbell, D. T., & Stanley, J. C. (1963). Experimental and quasi-experimental designs for generalized causal inference. Houghton Mifflin.

Campbell, D. T., Stanley, J. C., & Gage, N. L. (1966). *Experimental and quasi-experimental designs for research*. Rand McNally.

Campbell, M. J., & Machin, D. (1993). *Medical statistics: A commonsense approach*. Wiley.

Campbell, M. K., Elbourne, D. R., & Altman, D. G. (2004). CONSORT statement: Extension to cluster randomised trials. *British Medical Journal, 328*(7441), 702–708. https://doi.org/10.1136/bmj.328.7441.702

Canavos, G., & Koutrouvelis, J. (2008). *Introduction to the design & analysis of experiments*. Prentice Hall.

Cancer Research UK. (2019, February 13). *Phases of clinical trials*. www.cancerresearchuk.org/about-cancer/find-a-clinical-trial/what-clinical-trials-are/phases-of-clinical-trials

Card, N. A. (2015). *Applied meta-analysis for social science research*. Guilford Press.

Cardwell, S. M., Mazerolle, L., & Piquero, A. R. (2019). Truancy intervention and violent offending: Evidence from a randomised controlled trial. *Aggression and Violent Behavior, 49*, Article 101308. https://doi.org/10.1016/j.avb.2019.07.003

Carr, R., Slothower, M., & Parkinson, J. (2017). Do gang injunctions reduce violent crime? Four tests in Merseyside, UK. *Cambridge Journal of Evidence-Based Policing, 1*(4), 195–210. https://doi.org/10.1007/s41887-017-0015-x

Cartwright, N. (2004). Causation: One word, many things. *Philosophy of Science, 71*(5), 805–819. https://doi.org/10.1086/426771

Cartwright, N., & Hardie, J. (2012). *Evidence-based policy: A practical guide to doing it better.* Oxford University Press. https://doi.org/10.1093/acprof:os obl/9780199841608.001.0001

Casadevall, A., & Fang, F. C. (2010). Reproducible science. *Infection and Immunity, 78*(12), 4972–4975. https://doi.org/10.1128/IAI.00908-10

Casini, L. (2012). Causation: Many words, one thing? *THEORIA, 27*(2), 203–219. https://doi.org/10.1387/theoria.4067

Chalmers, I. (2001). Comparing like with like: Some historical milestones in the evolution of methods to create unbiased comparison groups in therapeutic experiments. *International Journal of Epidemiology, 30*(5), 1156–1164. https://doi.org/10.1093/ije/30.5.1156

Cheng, S. Y., Davis, M., Jonson-Reid, M., & Yaeger, L. (2019). Compared to what? A meta-analysis of batterer intervention studies using nontreated controls or comparisons. *Trauma, Violence, & Abuse.* Advance online publication. https://doi.org/10.1177/1524838019865927

Chivers, B., & Barnes, G. (2018). Sorry, wrong number: Tracking court attendance targeting through testing a "nudge" text. *Cambridge Journal of Evidence-Based Policing, 2*(1–2), 4–34. https://doi.org/10.1007/s41887-018-0023-5

Chow, S.-C., & Liu, J.-P. (2004). *Design and analysis of clinical trials: Concepts and methodologies.* Wiley-IEEE.

Chu, R., Walter, S. D., Guyatt, G., Devereaux, P. J., Walsh, M., Thorlund, K., & Thabane, L. (2012). Assessment and implication of prognostic imbalance in randomised controlled trials with a binary outcome: A simulation study. *PLOS ONE, 7*(5), Article e36677. https://doi.org/10.1371/journal.pone.0036677

Clarke, R. V. G., & Cornish, D. B. (1972). *The controlled trial in institutional research: Paradigm or pitfall for penal evaluators?* (Home Office Research Studies No. 15). Her Majesty's Stationery Office. http://library.college.police.uk/docs/hors/hors15.pdf

Cochran, W. G., & Cox, G. M. (1957). *Experimental designs.* Wiley.

Cohen, J. (2013). *Statistical power analysis for the behavioral sciences.* Academic Press. https://doi.org/10.4324/9780203771587

Coldwell, D., & Herbst, F. (2004). *Business research*. Juta.

Congdon, W. J., Kling, J. R., Ludwig, J., & Mullainathan, S. (2017). Social policy: Mechanism experiments and policy evaluations. In A. Banerjee & E. Duflo (Eds.), *Handbook of economic field experiments* (Vol. 2, pp. 389–426). North-Holland.

Conover, W. J. (1999). *Practical nonparametric statistics* (3rd ed.). Wiley.

Cook, T. D., & Campbell, D. T. (1979). The design and conduct of true experiments and quasi-experiments in field settings. In R. T. Mowday & R. M. Steers (Eds.), *Research in organizations: Issues and controversies* (pp. 223–326). Goodyear.

Cowen, N., & Cartwright, N. (2019). Street-level theories of change: Adapting the medical model of evidence-based practice for policing. In N. Fielding, K. Bullock, & S. Holdaway (Eds.), *Critical reflections on evidence-based policing* (pp. 52–71). Routledge. https://doi.org/10.4324/9780429488153-4

Cox, D. R. (1958). *The planning of experiments*. Wiley.

Cox, D. R. (1972). Regression models and life-tables. *Journal of the Royal Statistical Society: Series B (Methodological)*, *34*(2), 187–202. https://doi.org/10.1111/j.2517-6161.1972.tb00899.x

Cox, D. R., & Reid, N. (2000). *The theory of the design of experiments*. Chapman & Hall/CRC. https://doi.org/10.1201/9781420035834

Crandall, M., Eastman, A., Violano, P., Greene, W., Allen, S., Block, E., Christmas, A. B., Dennis, A., Duncan, T., Foster, S., Goldberg, S., Hirsh, M., Joseph, D., Lommel, K., Pappas, P., & Shillinglaw, W. (2016). Prevention of firearm-related injuries with restrictive licensing and concealed carry laws: An Eastern Association for the Surgery of Trauma systematic review. *Journal of Trauma and Acute Care Surgery*, *81*(5), 952–960. https://doi.org/10.1097/TA.0000000000001251

Crolley, J., Roys, D., Thyer, B. A., & Bordnick, P. S. (1998). Evaluating outpatient behavior therapy of sex offenders: A pretest-posttest study. *Behavior Modification*, *22*(4), 485–501. https://doi.org/10.1177/01454455980224003

Cumberbatch, J. R., & Barnes, G. C. (2018). This nudge was not enough: A randomised trial of text message reminders of court dates to victims and witnesses. *Cambridge Journal of Evidence-Based Policing*, *2*(1–2), 35–51. https://doi.org/10.1007/s41887-018-0024-4

Daly, J. E., & Pelowski, S. (2000). Predictors of dropout among men who batter: A review of studies with implications for research and practice. *Violence and Victims*, *15*(2), 137–160. https://doi.org/10.1891/0886-6708.15.2.137

Damen, H. (2017). Learning from implementing: A case study of a violence reduction programme for a night time economy area (England) [Unpublished master's thesis]. Institute of Criminology, University of Cambridge.

Danziger, S., Levav, J., & Avnaim-Pesso, L. (2011). Extraneous factors in judicial decisions. *Proceedings of the National Academy of Sciences of the United States of America, 108*(17), 6889–6892. https://doi.org/10.1073/pnas.1018033108

Davies, P. (1999). What is evidence-based education? *British Journal of Educational Studies, 47*(2), 108–121. https://doi.org/10.1111/1467-8527.00106

Davies, P., & Francis, P. (2018). *Doing criminological research.* Sage.

Davis, R. C., Weisburd, D., & Hamilton, E. E. (2010). Preventing repeat incidents of family violence: A randomised field test of a second responder program. *Journal of Experimental Criminology, 6*(4), 397–418. https://doi.org/10.1007/s11292-010-9107-3

Dawid, A. P. (2000). Causal inference without counterfactuals. *Journal of the American Statistical Association, 95*(450), 407–424. https://doi.org/10.1080/01621459.2000.10474210

Dawson, T. E. (1997). *A primer on experimental and quasi-experimental design* (ED406440). ERIC. https://files.eric.ed.gov/fulltext/ED406440.pdf

Day, S. J., & Altman, D. G. (2000). Blinding in clinical trials and other studies. *British Medical Journal, 321*(7259), 504. https://doi.org/10.1136/bmj.321.7259.504

DeAngelo, G., Toger, M., & Weisburd, S. (2020). *Police response times and injury outcomes* (CEPR Discussion Paper No. DP14536). SSRN. https://ssrn.com/abstract=3594157

De Boer, M. R., Waterlander, W. E., Kuijper, L. D., Steenhuis, I. H., & Twisk, J. W. (2015). Testing for baseline differences in randomised controlled trials: An unhealthy research behavior that is hard to eradicate. *International Journal of Behavioral Nutrition and Physical Activity, 12*(1), Article 4. https://doi.org/10.1186/s12966-015-0162-z

De Brito, C., & Ariel, B. (2017). Does tracking and feedback boost patrol time in hot spots? Two tests. *Cambridge Journal of Evidence-Based Policing, 1*(4), 244–262. https://doi.org/10.1007/s41887-017-0018-7

Delaney, C. (2006). *The effects of focused deterrence on gang homicide: An evaluation of Rochester's ceasefire program* [Master's thesis, Rochester Institute of Technology]. RIT Scholar Works. https://scholarworks.rit.edu/cgi/viewcontent.cgi?article=8208&context=theses

Denley, J., & Ariel, B. (2019). Whom should we target to prevent? Analysis of organized crime in England using intelligence records. *European Journal of Crime, Criminal Law and Criminal Justice, 27*(1), 13–44. https://doi.org/10.1163/15718174-02701003

Devereaux, P. J., Bhandari, M., Clarke, M., Montori, V. M., Cook, D. J., Yusuf, S., Sackett, D. L., Cina, C. S., Walter, S. D., Haynes, B., Schunemann, H. J.,

Norman, G. R., & Guyatt, G. H. (2005). Need for expertise based randomised controlled trials. *British Medical Journal, 330*(7482), 330–388. https://doi.org/10.1136/bmj.330.7482.88

De Winter, J. C. (2013). Using the student's *t*-test with extremely small sample sizes. *Practical Assessment, Research, and Evaluation, 18*, Article 10.

Dezember, A., Stoltz, M., & Marmolejo, L. (2020). The lack of experimental research in criminology: Evidence from Criminology and Justice Quarterly. *Journal of Experimental Criminology*. Advance online publication. https://doi.org/10.1007/s11292-020-09425-y

Dittmann, M. (2004). What makes good people do bad things. *Monitor on Psychology, 35*(9), 68. https://doi.org/10.1037/e309182005-051

Donner, A., & Klar, N. (2010). *Design and analysis of cluster randomisation trials in health research*. Arnold.

Drezner, D. W. (2020). *The toddler in chief: What Donald Trump teaches us about the modern presidency*. University of Chicago Press. https://doi.org/10.7208/chicago/9780226714394.001.0001

Drover, P., & Ariel, B. (2015). Leading an experiment in police body-worn video cameras. *International Criminal Justice Review, 25*(1), 80–97. https://doi.org/10.1177/1057567715574374

Duckett, S., & Griffiths, K. (2016). *Perils of place: Identifying hotspots of health inequalities* (Report No. 2016-10). Grattan Institute. https://grattan.edu.au/wp-content/uploads/2016/07/874-Perils-of-Place.pdf

Duckworth, A. L., & Kern, M. L. (2011). A meta-analysis of the convergent validity of self-control measures. *Journal of Research in Personality, 45*(3), 259–268. https://doi.org/10.1016/j.jrp.2011.02.004

Dudfield, G., Angel, C., Sherman, L. W., & Torrence, S. (2017). The "power curve" of victim harm: Targeting the distribution of crime harm index values across all victims and repeat victims over 1 year. *Cambridge Journal of Evidence-Based Policing, 1*(1), 38–58. https://doi.org/10.1007/s41887-017-0001-3

Dukes, R. L., Ullman, J. B., & Stein, J. A. (1995). An evaluation of DARE (Drug Abuse Resistance Education), using a Solomon four-group design with latent variables. *Evaluation Review, 19*(4), 409–435. https://doi.org/10.1177/0193841X9501900404

Dulachan, D. (2014). *Tracking citizens' complaints against police in Trinidad (Trinidad and Tobago)* [Unpublished master's thesis]. Institute of Criminology, University of Cambridge.

Eby, L. T., Allen, T. D., Evans, S. C., Ng, T., & DuBois, D. L. (2008). Does mentoring matter? A multidisciplinary meta-analysis comparing mentored and non-mentored individuals. *Journal of Vocational Behavior, 72*(2), 254–267. https://doi.org/10.1016/j.jvb.2007.04.005

Eck, J. E., & Weisburd, D. (Eds.). (1995). *Crime and place* (Vol. 4). Criminal Justice Press.

Efron, B. (1971). Forcing a sequential experiment to be balanced. *Biometrika, 58*(3), 403–417. https://doi.org/10.1093/biomet/58.3.403

Ellis, S., & Arieli, S. (1999). Predicting intentions to report administrative and disciplinary infractions: Applying the reasoned action model. *Human Relations, 52*(7), 947–967. https://doi.org/10.1177/001872679905200705

Elvik, R. (2016). Association between increase in fixed penalties and road safety outcomes: A meta-analysis. *Accident Analysis & Prevention, 92*, 202–210. https://doi.org/10.1016/j.aap.2016.03.028

Engel, R. J., & Schutt, R. K. (2014). *Fundamentals of social work research.* Sage.

Englefield, A., & Ariel, B. (2017). Searching for influencing actors in co-offending networks: The recruiter. *International Journal of Social Science Studies, 5*(5), 24–45. https://doi.org/10.11114/ijsss.v5i5.2351

Fagerland, M. W. (2012). T-tests, non-parametric tests, and large studies: A paradox of statistical practice? *BMC Medical Research Methodology, 12*(1), Article 78. https://doi.org/10.1186/1471-2288-12-78

Farrington, D. P. (1983). Randomised experiments on crime and justice. *Crime and Justice, 4*, 257–308. https://doi.org/10.1086/449091

Farrington, D. P. (1986). Age and crime. *Crime and Justice, 7*, 189–250. https://doi.org/10.1086/449114

Farrington, D. P. (2003a). British randomised experiments on crime and justice. *Annals of the American Academy of Political and Social Science, 589*(1), 150–167. https://doi.org/10.1177/0002716203254695

Farrington, D. P. (2003b). A short history of randomized experiments in criminology. *Evaluation Review, 27*(3), 218–227. https://journals.sagepub.com/doi/pdf/10.1177/0193841X03027003002

Farrington, D. P. (2006). Developmental criminology and risk-focused prevention. In M. Maguire, R. Morgan, & R. Reiner (Eds.), *The Oxford handbook of criminology* (pp. 657–701). Oxford University Press.

Farrington, D. P., Gottfredson, D. C., Sherman, L. W., & Welsh, B. C. (2002). The Maryland scientific methods scale. In D. P. Farrington, D. L. MacKenzie, L. W. Sherman, & B. C. Welsh (Eds.), *Evidence-based crime prevention* (pp. 13–21). Routledge. https://doi.org/10.4324/9780203166697_chapter_2

Farrington, D. P., & Petrosino, A. (2001). The Campbell collaboration crime and justice group. *Annals of the American Academy of Political and Social Science, 578*(1), 35–49. https://doi.org/10.1177/000271620157800103

Farrington, D. P., & Welsh, B. C. (2005). Randomised experiments in criminology: What have we learned in the last two decades? *Journal of Experimental Criminology, 1*(1), 9–38. https://doi.org/10.1007/s11292-004-6460-0

Farrington, D. P., & Welsh, B. C. (2013). Randomised experiments in criminology: What has been learned from long-term follow-ups. In B. C. Welsh, A. A. Braga, & G. J. N. Bruinsma (Eds.), *Experimental criminology: Prospects for advancing science and public policy* (pp. 111–140). Cambridge University Press. https://doi.org/10.1017/CBO9781139424776.010

Farrohknia, N., Castrén, M., Ehrenberg, A., Lind, L., Oredsson, S., Jonsson, H., Asplund, K., & Göransson, K. E. (2011). Emergency department triage scales and their components: A systematic review of the scientific evidence. *Scandinavian Journal of Trauma, Resuscitation and Emergency Medicine, 19*, Article 42. https://doi.org/10.1186/1757-7241-19-42

Feder, L., & Boruch, R. F. (2000). The need for experiments in criminal justice settings. *Crime & Delinquency, 46*(3), 291–294. https://doi.org/10.1177/0011128700046003001

Feder, L., Niolon, P. H., Campbell, J., Wallinder, J., Nelson, R., & Larrouy, H. (2011). The need for experimental methodology in intimate partner violence: Finding programs that effectively prevent IPV. *Violence Against Women, 17*(3), 341–360. https://doi.org/10.1177/1077801211398620

Feder, L., & Wilson, D. B. (2005). A meta-analytic review of court-mandated batterer intervention programs: Can courts affect abusers' behavior? *Journal of Experimental Criminology, 1*(2), 239–262. https://doi.org/10.1007/s11292-005-1179-0

Fergusson, D., Aaron, S. D., Guyatt, G., & Hébert, P. (2002). Post-randomisation exclusions: The intention to treat principle and excluding patients from analysis. *British Medical Journal, 325*(7365), 652–654. https://doi.org/10.1136/bmj.325.7365.652

Festing, M. F., & Altman, D. G. (2002). Guidelines for the design and statistical analysis of experiments using laboratory animals. *ILAR Journal, 43*(4), 244–258. https://doi.org/10.1093/ilar.43.4.244

Fewell, Z., Davey Smith, G., & Sterne, J. A. (2007). The impact of residual and unmeasured confounding in epidemiologic studies: A simulation study. *American Journal of Epidemiology, 166*(6), 646–655. https://doi.org/10.1093/aje/kwm165

Fisher, R. A. (1925). *Statistical methods for research workers*. Oliver & Boyd.

Fisher, R. A. (1935). *The design of experiments*. Oliver & Boyd.

Fitzgerald, J. D., & Cox, S. M. (1994). *Research methods in criminal justice: An introduction* (2nd ed.). Nelson-Hall.

Fitz-Gibbon, C. T. (1999). Education: High potential not yet realized. *Public Money & Management, 19*(1), 33–40. https://doi.org/10.1111/1467-9302.00150

Fixsen, D. L., Naoom, S. F., Blase, K. A., Friedman, R. M., & Wallace, F. (2005). *Implementation research: A synthesis of the literature* (FMHI Publication No. 231).

National Implementation Research Network, Louis de la Parte Florida Mental Health Institute, University of South Florida. https://nirn.fpg.unc.edu/sites/nirn.fpg.unc.edu/files/resources/NIRN-MonographFull-01-2005.pdf

Fonow, M. M., Richardson, L., & Wemmerus, V. A. (1992). Feminist rape education: Does it work? *Gender & Society, 6*(1), 108–121. https://doi.org/10.1177/089124392006001007

Food and Drug Administration (FDA). (2018, October). *Master protocols: Efficient clinical trial design strategies to expedite development of oncology drugs and biologics: Draft guidance for industry.* www.fda.gov/regulatory-information/search-fda-guidance-documents/master-protocols-efficient-clinical-trial-design-strategies-expedite-development-oncology-drugs-and

Forsyth, D. R. (2018). *Group dynamics.* Cengage Learning.

Freedman, L. P., Venugopalan, G., & Wisman, R. (2017). Reproducibility2020: Progress and priorities. *F1000Research, 6,* Article 604. https://doi.org/10.12688/f1000research.11334.1

Friedman, L. M., Furberg, C. D., & DeMets, D. L. (1985). *Fundamentals in clinical trials* (2nd ed.). PSG.

Frydensberg, C., Ariel, B., & Bland, M. (2019). Targeting the most harmful co-offenders in Denmark: A social network analysis approach. *Cambridge Journal of Evidence-Based Policing, 3*(1–2), 21–36. https://doi.org/10.1007/s41887-019-00035-x

Gacula, M. (2005). *Design and analysis of sensory optimization.* Blackwell. https://doi.org/10.1002/9780470385012

Galton, F. (1869). *Hereditary genius: An inquiry into its laws and consequences* (Vol. *27*). Macmillan. https://doi.org/10.1037/13474-000

Garner, J. (1990, November 7–10). Two, three . . . many experiments. The use and meaning of replication in social science research [Paper presentation]. Annual meeting of the American Society of Criminology, Baltimore, MD, US.

Garner, J. H., & Visher, C. A. (2003). The production of criminological experiments. *Evaluation Review, 27*(3), 316–335. https://doi.org/10.1177/0193841X03027003006

Gazal-Ayal, O., & Sulitzeanu-Kenan, R. (2010). Let my people go: Ethnic in-group bias in judicial decisions – evidence from a randomised natural experiment. *Journal of Empirical Legal Studies, 7*(3), 403–428. https://doi.org/10.1111/j.1740-1461.2010.01183.x

Gelman, A., & Loken, E. (2013). The garden of forking paths: Why multiple comparisons can be a problem, even when there is no "fishing expedition" or "p-hacking" and the research hypothesis was posited ahead of time. Department of Statistics, Columbia University. www.stat.columbia.edu/~gelman/research/unpublished/p_hacking.pdf

Gelman, A., Skardhamar, T., & Aaltonen, M. (2020). Type M error might explain Weisburd's paradox. *Journal of Quantitative Criminology*, *36*(2), 295–304. https://doi.org/10.1007/s10940-017-9374-5

Gerber, A. S., & Green, D. P. (2011, July). Field experiments and natural experiments. In R. E. Goodin (Ed.), *The Oxford handbook of political science* (pp. 1108–1132). Oxford University Press. https://doi.org/10.1093/oxfordhb/9780199604456.013.0050

Gesch, C. B., Hammond, S. M., Hampson, S. E., Eves, A., & Crowder, M. J. (2002). Influence of supplementary vitamins, minerals and essential fatty acids on the antisocial behaviour of young adult prisoners: Randomised, placebo-controlled trial. *British Journal of Psychiatry*, *181*(1), 22–28. https://doi.org/10.1192/bjp.181.1.22

Gibaldi, M., & Sullivan, S. (1997). Intention-to-treat analysis in randomized trials: Who gets counted? *Journal of Clinical Pharmacology*, *37*(8), 667–672. https://doi.org/10.1002/j.1552-4604.1997.tb04353.x

Gilbert, D. T., King, G., Pettigrew, S., & Wilson, T. D. (2016). Comment on "estimating the reproducibility of psychological science." *Science*, *351*(6277), 1037–1037. https://doi.org/10.1126/science.aad7243

Gill, C. E. (2011). Missing links: How descriptive validity impacts the policy relevance of randomised controlled trials in criminology. *Journal of Experimental Criminology*, *7*(3), Article 201. https://doi.org/10.1007/s11292-011-9122-z

Gill, C. E., & Weisburd, D. (2013). Increasing equivalence in small-sample place-based experiments: Taking advantage of block randomisation methods. In B. Welsh, A. Braga, & G. Bruinsma (Eds.), *Experimental criminology: Prospects for advancing science and public policy* (pp. 141–162). Cambridge University Press. https://doi.org/10.1017/CBO9781139424776.011

Gill, J. L. (1984). Heterogeneity of variance in randomised block experiments. *Journal of Animal Science*, *59*(5), 1339–1344. https://doi.org/10.2527/jas1984.5951339x

Giraldo, O., Garcia, A., & Corcho, O. (2018). A guideline for reporting experimental protocols in life sciences. *PeerJ*, *6*, Article e4795. https://doi.org/10.7717/peerj.4795

Glass, G. V. (1976). Primary, secondary, and meta-analysis of research. *Educational Researcher*, *5*(10), 3–8. https://doi.org/10.3102/0013189X005010003

Goldacre, B. (2014). *Bad pharma: How drug companies mislead doctors and harm patients*. Macmillan.

Gondolf, E. W. (2009a). Implementing mental health treatment for batterer program participants: Interagency breakdowns and underlying issues. *Violence Against Women*, *15*(6), 638–655. https://doi.org/10.1177/1077801209332189

Gondolf, E. W. (2009b). Outcomes from referring batterer program participants to mental health treatment. *Journal of Family Violence, 24*(8), Article 577. https://doi.org/10.1007/s10896-009-9256-1

Gordon, M. S., Kinlock, T. W., Schwartz, R. P., & O'Grady, K. E. (2008). A randomised clinical trial of methadone maintenance for prisoners: Findings at 6 months post-release. *Addiction, 103*(8), 1333–1342. https://doi.org/10.1111/j.1360-0443.2008.002238.x

Gottfredson, M., & Hirschi, T. (1987). The methodological adequacy of longitudinal research on crime. *Criminology, 25*(3), 581–614. https://doi.org/10.1111/j.1745-9125.1987.tb00812.x

Gottfredson, M. R., & Hirschi, T. (1990). *A general theory of crime.* Stanford University Press.

Gravel, J. (2007). The intention-to-treat approach in randomised controlled trials: Are authors saying what they do and doing what they say? *Clinical Trials, 4*(4), 350–356. https://doi.org/10.1177/1740774507081223

Green, S., & Byar, D. (1978). The effect of stratified randomisation on size and power of statistical tests in clinical trials. *Journal of Chronic Diseases, 31*(6–7), 445–454. https://doi.org/10.1016/0021-9681(78)90008-5

Greene, J. A. (1999). Zero tolerance: A case study of police policies and practices in New York City. *Crime & Delinquency, 45*(2), 171–187. https://doi.org/10.1177/0011128799045002001

Greene, J. R. (2010). Collaborations between police and research/academic organisations: Some prescriptions from the field. In J. Klofas, N. Hipple, & E. McGarrell (Eds.), *The new criminal justice* (pp. 121–127). Routledge.

Gresswell, M. (2018). *Tracking the use of body-worn video in Bedfordshire police in 2017 (England)* [Unpublished master's thesis]. Institute of Criminology, University of Cambridge.

Grimshaw, J., Campbell, M., Eccles, M., & Steen, N. (2000). Experimental and quasi-experimental designs for evaluating guideline implementation strategies. *Family Practice, 17*(Suppl. 1), S11–S16. https://doi.org/10.1093/fampra/17.suppl_1.S11

Grol, R. (2001). Successes and failures in the implementation of evidence-based guidelines for clinical practice. *Medical care, 39*(8), II-46–II-54. https://doi.org/10.1097/00005650-200108002-00003

Grommon, E., Rydberg, J., & Bynum, T. (2012). *Understanding the challenges facing offenders upon their return to the community: Final report.* Michigan Justice Statistics Center. https://cj.msu.edu/_assets/pdfs/mjsc/MJSC-UCFOURC-Jan2012.pdf

Grommon, E., Rydberg, J., & Carter, J. G. (2017). Does GPS supervision of intimate partner violence defendants reduce pretrial misconduct? Evidence from a quasi-experimental study. *Journal of Experimental Criminology, 13*(4), 483–504. https://doi.org/10.1007/s11292-017-9304-4

Grossmith, L., Owens, C., Finn, W., Mann, D., Davies, T., & Baika, L. (2015). *Police, camera, evidence: London's cluster randomised controlled trial of body worn video.* College of Policing. https://bja.ojp.gov/sites/g/files/xyckuh186/files/bwc/pdfs/CoPBWVreportNov2015.pdf

Guo, Y., Kopec, J. A., Cibere, J., Li, L. C., & Goldsmith, C. H. (2016). Population survey features and response rates: A randomised experiment. *American Journal of Public Health, 106*(8), 1422–1426. https://doi.org/10.2105/AJPH.2016.303198

Haberman, C. P., Clutter, J. E., & Henderson, S. (2018). A quasi-experimental evaluation of the impact of bike-sharing stations on micro-level robbery occurrence. *Journal of Experimental Criminology, 14*(2), 227–240. https://doi.org/10.1007/s11292-017-9312-4

Hadorn, D. C., Baker, D., Hodges, J. S., & Hicks, N. (1996). Rating the quality of evidence for clinical practice guidelines. *Journal of Clinical Epidemiology, 49*(7), 749–754. https://doi.org/10.1016/0895-4356(96)00019-4

Hallstrom, A., & Davis, K. (1988). Imbalance in treatment assignments in stratified blocked randomisation. *Controlled Clinical Trials, 9*(4), 375–382. https://doi.org/10.1016/0197-2456(88)90050-5

Handelsman, J., Ebert-May, D., Beichner, R., Bruns, P., Chang, A., DeHaan, R., Gentile, J., Lauffer, S., Stewart, J., Tilghman, S. M., & Wood, W. B. (2004). Scientific teaching. *Science, 304*(5670), 521–522. https://doi.org/10.1126/science.1096022

Hare, A. P. (1976). *Handbook of small group research.* Free Press.

Harris, W. S., Gowda, M., Kolb, J. W., Strychacz, C. P., Vacek, J. L., Jones, P. G., Forker, A., O'Keefe, J. H., & McCallister, B. D. (1999). A randomized, controlled trial of the effects of remote, intercessory prayer on outcomes in patients admitted to the coronary care unit. *Archives of Internal Medicine, 159*(19), 2273–2278. https://doi.org/10.1001/archinte.159.19.2273

Harrison, G. W., & List, J. A. (2004). Field experiments. *Journal of Economic Literature, 42*(4), 1009–1055. https://doi.org/10.1257/0022051043004577

Hasisi, B., Shoham, E., Weisburd, D., Haviv, N., & Zelig, A. (2016). The "care package," prison domestic violence programs and recidivism: A quasi-experimental study. *Journal of Experimental Criminology, 12*(4), 563–586. https://doi.org/10.1007/s11292-016-9266-y

Haviland, A. M., & Nagin, D. S. (2005). Causal inference with group-based trajectory models. *Psychometrika, 70*(3), 557–578. https://doi.org/10.1007/s11336-004-1261-y

Haviland, A. M., Nagin, D. S., & Rosenbaum, P. R. (2007). Combining propensity score matching and group-based trajectory analysis in an observational study. *Psychological Methods, 12*(3), 247–267. https://doi.org/10.1037/1082-989X.12.3.247

Haviv, N., Weisburd, D., Hasisi, B., Shoham, E., & Wolfowicz, M. (2019). Do religious programs in prison work? A quasi-experimental evaluation in the Israeli prison service. *Journal of Experimental Criminology*. Advance online publication. https://doi.org/10.1007/s11292-019-09375-0

Hayden, C., & Jenkins, C. (2014). "Troubled families" programme in England: "Wicked problems" and policy-based evidence. *Policy Studies, 35*(6), 631–649. https://doi.org/10.1080/01442872.2014.971732

Hayes, A. (2020, March 10). Error term. *Investopedia*. Retrieved August 22, 2020, from https://www.investopedia.com/terms/e/errorterm.asp#:~:text=An%20error%20term%20is%20a,variables%20and%20the%20dependent%20variables.

Hayward, G. D., Steinhorst, R. K., & Hayward, P. H. (1992). Monitoring boreal owl populations with nest boxes: Sample size and cost. *Journal of Wildlife Management, 56*(4), 777–785. https://doi.org/10.2307/3809473

Head, M. L., Holman, L., Lanfear, R., Kahn, A. T., & Jennions, M. D. (2015). The extent and consequences of *p*-hacking in science. *PLOS Biology, 13*(3), Article e1002106. https://doi.org/10.1371/journal.pbio.1002106

Heckert, D. A., & Gondolf, E. W. (2005). Do multiple outcomes and conditional factors improve prediction of batterer reassault? *Violence and Victims, 20*(1), 3–24. https://doi.org/10.1891/vivi.2005.20.1.3

Heckman, J. J. (1979). Sample selection bias as a specification error. *Econometrica, 47*(1), 153–161. https://doi.org/10.2307/1912352

Heckman, J. J. (1990). Varieties of selection bias. *American Economic Review, 80*(2), 313–318.

Heckman, J. J., Ichimura, H., Smith, J., & Todd, P. (1998). Characterizing selection bias using experimental data. *Econometrica, 66*(5), 1017–1098. https://doi.org/10.2307/2999630

Hedström, P. (2005). *Dissecting the social*. Cambridge University Press. https://doi.org/10.1017/CBO9780511488801

Hedström, P., Swedberg, R., & Hernes, G. (Eds.). (1998). *Social mechanisms: An analytical approach to social theory*. Cambridge University Press. https://doi.org/10.1017/CBO9780511663901

Heinich, R. (1970). *Technology and the management of instruction* (Monograph No. 4). Association for Educational Communications and Technology.

Heinsman, D. T., & Shadish, W. R. (1996). Assignment methods in experimentation: When do nonrandomised experiments approximate answers from randomised experiments? *Psychological Methods, 1*(2), 154–169. https://doi.org/10.1037/1082-989X.1.2.154

Henderson, R. (2014). *Learning the lessons of implementation: A case study of an RCT in body-worn video (Northern Ireland)* [Unpublished master's thesis]. Institute of Criminology, University of Cambridge.

Henstock, D., & Ariel, B. (2017). Testing the effects of police body-worn cameras on use of force during arrests: A randomised controlled trial in a large British police force. *European Journal of Criminology, 14*(6), 720–750. https://doi.org/10.1177/1477370816686120

Heukelom, F. (2009). *Origin and interpretation of internal and external validity in economics* (NiCE Working Paper No. 09-111). Nijmegen Center for Economics, Institute for Management Research, Radboud University. https://repository.ubn.ru.nl/bitstream/handle/2066/74891/74891.pdf

Higginson, A., Eggins, E., Mazerolle, L., & Stanko, E. (2014). *The global policing database* [Database and protocol]. https://gpd.uq.edu.au/s/gpd/page/about

Hill, A. B. (1951). The clinical trial. *British Medical Bulletin, 7*(4), 278–282. https://doi.org/10.1093/oxfordjournals.bmb.a073919

Hinkle, J. C., Weisburd, D., Famega, C., & Ready, J. (2013). The problem is not just sample size: The consequences of low base rates in policing experiments in smaller cities. *Evaluation Review, 37*(3–4), 213–238. https://doi.org/10.1177/0193841X13519799

Hinkle, J. C., Weisburd, D., Telep, C. W., & Petersen, K. (2020). Problem-oriented policing for reducing crime and disorder: An updated systematic review and meta-analysis. *Campbell Systematic Reviews, 16*(2), Article e1089. https://doi.org/10.1002/cl2.1089

Hirschel, J. D., & Hutchison, I. W., III. (1992). Female spouse abuse and the police response: The Charlotte, North Carolina experiment. *Journal of Criminal Law and Criminology, 83*(1), 73–119. https://doi.org/10.2307/1143825

Hirschi, T., & Gottfredson, M. (1983). Age and the explanation of crime. *American Journal of Sociology, 89*(3), 552–584. https://doi.org/10.1086/227905

Hoffmann, T. C., Glasziou, P. P., Boutron, I., Milne, R., Perera, R., Moher, D., Altman, D. G., Barbour, V., Macdonald, H., Johnston, M., Lamb, S. E., Dixon-Woods, M., McCulloch, P., Wyatt, J. C., Chan, A. W., & Michie, S. (2014). Better reporting of interventions: Template for intervention description and replication (TIDieR) checklist and guide. *British Medical Journal, 348*, Article g1687. https://doi.org/10.1136/bmj.g1687

Højlund, F., & Ariel, B. (2019, July 10). Operation Knock Knock: Organised criminal group-recruit warnings in Copenhagen [Paper presentation]. The 12th international evidence-based policing conference, Cambridge, UK.

Holland, P. W. (1986). Statistics and causal inference. *Journal of the American Statistical Association, 81*(396), 945–960. https://doi.org/10.1080/01621459.1986.10478354

Hollis, S., & Campbell, F. (1999). What is meant by intention to treat analysis? Survey of published randomised controlled trials. *British Medical Journal*, *319*(7211), 670–674. https://doi.org/10.1136/bmj.319.7211.670

Hough, M., Bradford, B., Jackson, J., & Quinton, P. (2016). *Does legitimacy necessarily tame power? Some ethical issues in translating procedural justice principles into justice policy* (LSE Legal Studies Working Paper No. 13/2016). SSRN. https://doi.org/10.2139/ssrn.2783799

Høye, A. (2010). Are airbags a dangerous safety measure? A meta-analysis of the effects of frontal airbags on driver fatalities. *Accident Analysis & Prevention, 42*(6), 2030–2040. https://doi.org/10.1016/j.aap.2010.06.014

Høye, A. (2014). Speed cameras, section control, and kangaroo jumps: A meta-analysis. *Accident Analysis & Prevention, 73*, 200–208. https://doi.org/10.1016/j.aap.2014.09.001

Huey, L., & Mitchell, R. J. (2016). Unearthing hidden keys: Why pracademics are an invaluable (if underutilized) resource in policing research. *Policing, 10*(3), 300–307. https://doi.org/10.1093/police/paw029

Hunter, J. E., & Schmidt, F. L. (2004). *Methods of meta-analysis: Correcting error and bias in research findings*. Sage.

Irving, B., & Hilgendorf, L. (1980). *Police interrogation: A case study of current practice.* Her Majesty's Stationery Office.

Israel, T., Harkness, A., Delucio, K., Ledbetter, J. N., & Avellar, T. R. (2014). Evaluation of police training on LGBTQ issues: Knowledge, interpersonal apprehension, and self-efficacy. *Journal of Police and Criminal Psychology, 29*(2), 57–67. https://doi.org/10.1007/s11896-013-9132-z

Israel Police Service. (2019). *Annual report of Israel Police for 2018.* www.gov.il/BlobFolder/reports/police_annual_report_under_the_freedom_of_information_law_2018/he/annual_report_under_the_freedom_of_information_law_2018.pdf

Jacobs, J. B. (1989). *Drunk driving: An American dilemma.* University of Chicago Press. https://doi.org/10.7208/chicago/9780226222905.001.0001

Jaitman, L. (2018). *Frontiers in the economics of crime: Lessons for Latin America and the Caribbean* (Technical Note No. IDB-TN-01596). Office of Strategic Planning and Development Effectiveness, Inter-American Development Bank. https://doi.org/10.18235/0001482

Jansson, K. (2007). *British Crime Survey: Measuring crime for 25 years.* Home Office. https://webarchive.nationalarchives.gov.uk/20100408175022/http://www.homeoffice.gov.uk/rds/pdfs07/bcs25.pdf

Jaycox, L. H., McCaffrey, D., Eiseman, B., Aronoff, J., Shelley, G. A., Collins, R. L., & Marshall, G. N. (2006). Impact of a school-based dating violence prevention program among Latino teens: Randomised controlled effectiveness trial. *Journal of Adolescent Health, 39*(5), 694–704. https://doi.org/10.1016/j.jadohealth.2006.05.002

Jaynes, E. T. (2003). *Probability theory: The logic of science.* Cambridge University Press. https://doi.org/10.1017/CBO9780511790423

Jenkins, W. (2018). The effects of tracking and feedback on "officer in the case" compliance with victim updating requirements: A randomised controlled trial (England) [Unpublished master's thesis]. Institute of Criminology, University of Cambridge.

Jennings, W. G., Lynch, M. D., & Fridell, L. A. (2015). Evaluating the impact of police officer body-worn cameras (BWCs) on response-to-resistance and serious external complaints: Evidence from the Orlando police department (OPD) experience utilizing a randomised controlled experiment. *Journal of Criminal Justice, 43*(6), 480–486. https://doi.org/10.1016/j.jcrimjus.2015.10.003

Johnson, S. D., Davies, T., Murray, A., Ditta, P., Belur, J., & Bowers, K. (2017). Evaluation of operation swordfish: A near-repeat target-hardening strategy. *Journal of Experimental Criminology, 13*(4), 505–525. https://doi.org/10.1007/s11292-017-9301-7

Johnson, W. E. (1932). Probability: The deductive and inductive problems. *Mind, 41*(164), 409–423. https://doi.org/10.1093/mind/XLI.164.409

Jonathan-Zamir, T., Mastrofski, S. D., & Moyal, S. (2015). Measuring procedural justice in police-citizen encounters. *Justice Quarterly, 32*(5), 845–871. https://doi.org/10.1080/07418825.2013.845677

Jonathan-Zamir, T., Weisburd, D., Dayan, M., & Zisso, M. (2019). The proclivity to rely on professional experience and evidence-based policing: Findings from a survey of high-ranking officers in the Israel police. *Criminal Justice and Behavior, 46*(10), 1456–1474. https://doi.org/10.1177/0093854819842903

Jupp, V. R. (2006). *The Sage dictionary of social research methods.* Sage. https://doi.org/10.4135/9780857020116

Jupp, V. R. (2012). *Methods of criminological research.* Routledge. https://doi.org/10.4324/9780203423981

Kaiser, L. D. (2012). Dynamic randomisation and a randomisation model for clinical trials data. *Statistics in Medicine, 31*(29), 3858–3873. https://doi.org/10.1002/sim.5448

Kelley, T. (1927). *Interpretation of educational measurements.* World Books.

Kelling, G. L., Pate, T., Dieckman, D., & Brown, C. E. (1974). *The Kansas City preventive patrol experiment.* Police Foundation. www.policefoundation.org/wp-content/uploads/2015/07/Kelling-et-al.-1974-THE-KANSAS-CITY-PREVENTIVE-PATROL-EXPERIMENT.pdf

Kempthorne, O. (1952). *The design and analysis of experiments*. Wiley.

Kempthorne, O. (1955). The randomization theory of experimental inference. *Journal of the American Statistical Association, 50*(271), 946–967. https://doi.org/10.1080/01621459.1955.10501979

Kendall, J. (2003). Designing a research project: Randomised controlled trials and their principles. *Emergency Medicine Journal, 20*(2), 164–168. https://doi.org/10.1136/emj.20.2.164

Kenny, D. A., Kashy, D. A., & Cook, W. L. (2006). *Dyadic data analysis*. Guilford Press.

Kernan, W. N., Viscoli, C. M., Makuch, R. W., Brass, L. M., & Horwitz, R. I. (1999). Stratified randomisation for clinical trials. *Journal of Clinical Epidemiology, 52*(1), 19–26. https://doi.org/10.1016/S0895-4356(98)00138-3

Kessler, J., & Vesterlund, L. (2015). The external validity of laboratory experiments: The misleading emphasis on quantitative effects. In G. R. Fréchette & A. Schotte (Eds.), *Handbook of experimental economic methodology* (Vol. *18*, pp. 392–405). Oxford University Press. https://doi.org/10.1093/acprof:oso/9780195328325.003.0020

Kim, T. K. (2015). T test as a parametric statistic. *Korean Journal of Anesthesiology, 68*(6), 540–546. https://doi.org/10.4097/kjae.2015.68.6.540

Kim, Y., & Steiner, P. (2016). Quasi-experimental designs for causal inference. *Educational Psychologist, 51*(3–4), 395–405. https://doi.org/10.1080/00461520.2016.1207177

Kimberlee, R. (2015). What is social prescribing? *Advances in Social Sciences Research Journal, 2*(1), 102–110. https://doi.org/10.14738/assrj.21.808

Klein, G. A. (2011). *Streetlights and shadows: Searching for the keys to adaptive decision making*. MIT Press.

Knapp, T. R. (2016). Why is the one-group pretest–posttest design still used? *Clinical Nursing Research, 25*(5), 467–472. https://doi.org/10.1177/1054773816666280

Koo, T. K., & Li, M. Y. (2016). A guideline of selecting and reporting intraclass correlation coefficients for reliability research. *Journal of Chiropractic Medicine, 15*(2), 155–163. https://doi.org/10.1016/j.jcm.2016.02.012

Koper, C. S. (1995). Just enough police presence: Reducing crime and disorderly behavior by optimizing patrol time in crime hot spots. *Justice Quarterly, 12*(4), 649–672. https://doi.org/10.1080/07418829500096231

Koper, C. S., & Mayo-Wilson, E. (2012). Police strategies to reduce illegal possession and carrying of firearms: Effects on gun crime. *Campbell Systematic Reviews, 8*(1), 1–53. https://doi.org/10.4073/csr.2012.11

Kovalsky, S., Hasisi, B., Haviv, N., & Elisha, E. (2020). Can yoga overcome criminality? The impact of yoga on recidivism in Israeli prisons. *International*

Journal of Offender Therapy and Comparative Criminology, 64(13–14), 1461–1481. https://doi.org/10.1177/0306624X20911899

Kownacki, R. J., & Shadish, W. R. (1999). Does Alcoholics Anonymous work? The results from a meta-analysis of controlled experiments. *Substance Use & Misuse, 34*(13), 1897–1916. https://doi.org/10.3109/10826089909039431

Kruskal, W. (1988). Miracles and statistics: The casual assumption of independence. *Journal of the American Statistical Association, 83*(404), 929–940. https://doi.org/10.1080/01621459.1988.10478682

Labriola, M., Rempel, M., & Davis, R. C. (2008). Do batterer programs reduce recidivism? Results from a randomized trial in the Bronx. *Justice Quarterly, 25*(2), 252–282. https://doi.org/10.1080/07418820802024945

Lachin, J. M. (1988a). Properties of simple randomisation in clinical trials. *Controlled Clinical Trials, 9*(4), 312–326. https://doi.org/10.1016/0197-2456(88)90046-3

Lachin, J. M. (1988b). Statistical properties of randomisation in clinical trials. *Controlled Clinical Trials, 9*(4), 289–311. https://doi.org/10.1016/0197-2456(88)90045-1

Lachin, J. M., & Bautista, O. M. (1995). Stratified-adjusted versus unstratified assessment of sample size and power for analyses of proportions. In P. F. Thall (Ed.), *Recent advances in clinical trial design and analysis* (pp. 203–223). Kluwer. https://doi.org/10.1007/978-1-4615-2009-2_10

Lachin, J. M., Matts, J. P., & Wei, L. J. (1988). Randomisation in clinical trials: Conclusions and recommendations. *Controlled Clinical Trials, 9*(4), 365–374. https://doi.org/10.1016/0197-2456(88)90049-9

Lagakos, S. W., & Pocock, S. J. (1984). Randomisation and stratification in cancer clinical trials: An international survey. In M. E. Buyse, M. J. Staquet, & R. J. Sylvester (Eds.), *Cancer clinical trials, methods and practice* (pp. 276–286). Oxford University Press.

Lájer, K. (2007). Statistical tests as inappropriate tools for data analysis performed on non-random samples of plant communities. *Folia Geobotanicat, 42*(2), Article 115. https://doi.org/10.1007/s11292-005-3541-7

Landenberger, N. A., & Lipsey, M. W. (2005). The positive effects of cognitive–behavioral programs for offenders: A meta-analysis of factors associated with effective treatment. *Journal of Experimental Criminology, 1*(4), 451–476. https://doi.org/10.1007/s11292-005-3541-7

Landsberger, H. A. (1958). *Hawthorne revisited: Management and the worker, its critics, and developments in human relations in industry.* Cornell University.

Langley, B., Ariel, B., Tankebe, J., Sutherland, A., Beale, M., Factor, R., & Weinborn, C. (2020). A simple checklist, that is all it takes: A cluster randomised controlled field trial on improving the treatment of suspected terrorists by the police. *Journal of Experimental Criminology*. Advance online publication. https://doi.org/10.1007/s11292-020-09428-9

Lanovaz, M. J., & Rapp, J. T. (2016). Using single-case experiments to support evidence-based decisions: How much is enough? *Behavior Modification, 40*(3), 377–395. https://doi.org/10.1177/0145445515613584

Lee, L. K., Fleegler, E. W., Farrell, C., Avakame, E., Srinivasan, S., Hemenway, D., & Monuteaux, M. C. (2017). Firearm laws and firearm homicides: A systematic review. *JAMA Internal Medicine, 177*(1), 106–119. https://doi.org/10.1001/jamainternmed.2016.7051

Leeuw, F. L., & Schmeets, H. (2016). *Empirical legal research: A guidance book for lawyers, legislators and regulators*. Edward Elgar. https://doi.org/10.4337/9781782549413

Leigh, A. (2018). *Randomistas: How radical researchers are changing our world*. Yale University Press.

Leppink, J. (2019). *Statistical methods for experimental research in education and psychology*. Springer. https://doi.org/10.1007/978-3-030-21241-4

Levitt, S. D., & List, J. A. (2007a). On the generalizability of lab behaviour to the field. *Canadian Journal of Economics, 40*(2), 347–370. https://doi.org/10.1111/j.1365-2966.2007.00412.x

Levitt, S. D., & List, J. A. (2007b). What do laboratory experiments measuring social preferences reveal about the real world? *Journal of Economic Perspectives, 21*(2), 153–174. https://doi.org/10.1257/jep.21.2.153

Levitt, S. D., & List, J. A. (2011). Was there really a Hawthorne effect at the Hawthorne plant? An analysis of the original illumination experiments. *American Economic Journal: Applied Economics, 3*(1), 224–238. https://doi.org/10.1257/app.3.1.224

Lewin, K. (1948). *Resolving social conflicts; selected papers on group dynamics*. Harper.

Lewis, D. (1974). Causation. *Journal of Philosophy, 70*(17), 556–567. https://doi.org/10.2307/2025310

Lewis, D. (2013). *Counterfactuals*. Wiley.

Liggins, A., Ratcliffe, J. H., & Bland, M. (2019). Targeting the most harmful offenders for an English police agency: Continuity and change of membership in the "felonious few." *Cambridge Journal of Evidence-Based Policing, 3*(3–4), 80–96. https://doi.org/10.1007/s41887-019-00039-7

Linning, S. J., Bowers, K., & Eck, J. E. (2019). Understanding the time-course of an intervention's mechanisms: A framework for improving experiments and evaluations. *Journal of Experimental Criminology, 15*(4), 593–610. https://doi. org/10.1007/s11292-019-09367-0

Linton, B., & Ariel, B. (2020). Random assignment with a smile: How to love "TheRandomiser". *Cambridge Journal of Evidence-Based Policing,* 1–5.

Lipsey, M. W. (1990). *Design sensitivity: Statistical power for experimental research.* Sage.

Lipsey, M. W. (2002). Meta-analysis and program outcome evaluation. *Socialvetenskaplig tidskrift, 9*(23), 194–208.

Lipsey, M. W., & Wilson, D. B. (1993). The efficacy of psychological, educational, and behavioral treatment: Confirmation from meta-analysis. *American Psychologist, 48*(12), 1181–1209. https://doi.org/10.1037/0003-066X.48.12.1181

List, J. A. (2011). Why economists should conduct field experiments and 14 tips for pulling one off. *Journal of Economic Perspectives, 25*(3), 3–16. https://doi. org/10.1257/jep.25.3.3

Loeber, R., & Farrington, D. P. (2014). Age–crime curve. In R. Loeber, D. P. Farrington, G. Bruinsma, & D. Weisburd (Eds.), *Encyclopedia of criminology and criminal justice* (pp. 12–18). Springer. https://doi.org/10.1007/978-1-4614-5690-2_474

Loftin, C., McDowall, D., Wiersema, B., & Cottey, T. J. (1991). Effects of restrictive licensing of handguns on homicide and suicide in the District of Columbia. *New England Journal of Medicine, 325*(23), 1615–1620. https://doi.org/10.1056/NEJM199112053252305

Lohr, S. L. (2019). *Sampling: Design and analysis.* Chapman & Hall/CRC. https://doi. org/10.1201/9780429296284

Lösel, F. (2018). Evidence comes by replication, however needs differentiation: The reproducibility issue in science and its relevance for criminology. *Journal of Experimental Criminology, 14*(3), 257–278. https://doi.org/10.1007/s11292-017-9297-z

Luca, D. L. (2015). Do traffic tickets reduce motor vehicle accidents? Evidence from a natural experiment. *Journal of Policy Analysis and Management, 34*(1), 85–106. https://doi.org/10.1002/pam.21798

Lum, C., Koper, C. S., Wilson, D. B., Stoltz, M., Goodier, M., Eggins, E., Higginson, A., & Mazerolle, L. (2020). Body-worn cameras' effects on police officers and citizen behavior: A systematic review. *Campbell Systematic Reviews, 16*(3), Article e1112. https://doi.org/10.1002/cl2.1112

Lum, C., & Yang, S. M. (2005). Why do evaluation researchers in crime and justice choose non-experimental methods? *Journal of Experimental Criminology, 1*(2), 191–213. https://doi.org/10.1007/s11292-005-1619-x

Lundivian, R. J., McFarlane, P. T., & Scarpitti, F. R. (1976). Delinquency prevention: A description and assessment of projects reported in the professional literature. *Crime & Delinquency, 22*(3), 297–308. https://doi.org/10.1177/001112877602200303

Macbeth, E., & Ariel, B. (2019). Place-based statistical versus clinical predictions of crime hot spots and harm locations in Northern Ireland. *Justice Quarterly, 36*(1), 93–126. https://doi.org/10.1080/07418825.2017.1360379

MacQueen, S., & Bradford, B. (2015). Enhancing public trust and police legitimacy during road traffic encounters: Results from a randomised controlled trial in Scotland. *Journal of Experimental Criminology, 11*(3), 419–443. https://doi.org/10.1007/s11292-015-9240-0

Magnusson, M. M. (2020). Bridging the gaps by including the police officer perspective? A study of the design and implementation of an RCT in police practice and the impact of pracademic knowledge. *Policing, 14*(2), 438–455. https://doi.org/10.1093/police/pay022

Maltz, M. D., Gordon, A. C., McDowall, D., & McCleary, R. (1980). An artifact in pretest-posttest designs: How it can mistakenly make delinquency programs look effective. *Evaluation Review, 4*(2), 225–240. https://doi.org/10.1177/0193841X8000400204

Mark, M. M., & Lenz-Watson, A. L. (2011). Ethics and the conduct of randomized experiments and quasi-experiments in field settings. In A. T. Panter & S. K. Sterba (Eds.), *Handbook of ethics in quantitative methodology* (pp. 185–209). Routledge.

Marsden, E., & Torgerson, C. J. (2012). Single group, pre-and post-test research designs: Some methodological concerns. *Oxford Review of Education, 38*(5), 583–616. https://doi.org/10.1080/03054985.2012.731208

Martin, R. M., & Marcuse, F. L. (1958). Characteristics of volunteers and nonvolunteers in psychological experimentation. *Journal of Consulting Psychology, 22*(6), 475–479. https://doi.org/10.1037/h0041496

Martin, S., Downe, J., Grace, C., & Nutley, S. (2010). Validity, utilization and evidence-based policy: The development and impact of performance improvement regimes in local public services. *Evaluation, 16*(1), 31–42. https://doi.org/10.1177/1356389009350119

Maskaly, J., Donner, C., Jennings, W. G., Ariel, B., & Sutherland, A. (2017). The effects of body-worn cameras (BWCs) on police and citizen outcomes: A state-of-the-art review. *Policing, 40*(4), 672–688.

Matts, J. P., & Lachin, J. M. (1988). Properties of permuted-block randomisation in clinical trials. *Controlled Clinical Trials, 9*(4), 327–344. https://doi.org/10.1016/0197-2456(88)90047-5

Maxwell, S. E. (1993). Covariate imbalance and conditional size: Dependence on model-based adjustments. *Statistics in Medicine, 12*(2), 101–109.

Mayo, E. (2009). Hawthorne and the Western Electric Company. In R. Stillman II (Ed.), *Public administration: Concepts and cases* (pp. 149–158). Cengage Learning.

Mazerolle, L., Antrobus, E., Bennett, S., & Tyler, T. R. (2013). Shaping citizen perceptions of police legitimacy: A randomised field trial of procedural justice. *Criminology, 51*(1), 33–63. https://doi.org/10.1111/j.1745-9125.2012.00289.x

Mazerolle, L., Bennett, S., Antrobus, E., Cardwell, S. M., Eggins, E., & Piquero, A. R. (2019). Disrupting the pathway from truancy to delinquency: A randomised field trial test of the longitudinal impact of a school engagement program. *Journal of Quantitative Criminology, 35*(4), 663–689. https://doi.org/10.1007/s10940-018-9395-8

Mazerolle, L., Bennett, S., Davis, J., Sargeant, E., & Manning, M. (2013). Procedural justice and police legitimacy: A systematic review of the research evidence. *Journal of Experimental Criminology, 9*(3), 245–274.

Mazerolle, L., Eggins, E., Higginson, A., & Stanko, B. (2017). Evidence-based policing as a disruptive innovation: The global policing database as a disruption tool. In J. Knutsson & L. Tompson (Eds.), *Advances in evidence-based policing* (pp. 117–138). Routledge. https://doi.org/10.4324/9781315518299

Mazerolle, L. G., Price, J. F., & Roehl, J. (2000). Civil remedies and drug control: A randomised field trial in Oakland, California. *Evaluation Review, 24*(2), 212–241. https://doi.org/10.1177/0193841X0002400203

McCabe, J. E., Morreale, S. A., & Tahiliani, J. R. (2016). The pracademic and academic in criminal justice education: A qualitative analysis. *Police Forum, 26*(1), 1–12.

McCall, W. A. (1923). *How to experiment in education.* Macmillan. https://doi.org/10.1037/13551-000

McCold, P., & Wachtel, B. (1998). Restorative Policing Experiment. *The Bethlehem Pennsylvania Police Family Group Conferencing Project.* Community Service Foundation, Pipersville, PA. (Republished 2012; Wipf & Stock.)

McCullough, B. D., & Wilson, B. (2005). On the accuracy of statistical procedures in Microsoft Excel, 2003. *Computational Statistics & Data Analysis, 49*(4), 1244–1252. https://doi.org/10.1016/j.csda.2004.06.016

McDaniel, M. A., Anderson, J. L., Derbish, M. H., & Morrisette, N. (2007). Testing the testing effect in the classroom. *European Journal of Cognitive Psychology, 19*(4–5), 494–513. https://doi.org/10.1080/09541440701326154

Meinert, C. L. (2012). *Clinical trials: Design, conduct and analysis* (Vol. 39). Oxford University Press.

Meinert, C. L., & Tonascia, S. (1986). *Clinical trials: Design, conduct, and analysis.* Oxford University Press.

Meyer, M. N., Heck, P. R., Holtzman, G. S., Anderson, S. M., Cai, M., Watts, D. J., & Chabris, C. F. (2019). Objecting to experiments that compare two unobjectionable policies or treatments. Proceedings of the National Academy of Sciences May 2019, *116*(22), 10723–10728. doi: 10.1073/pnas.1820701116

Michel, C. (2017). Examining the influence of increased knowledge about white-collar crime on attitudes toward it in the undergraduate classroom. *Journal of Criminal Justice Education, 28*(1), 52–73. https://doi.org/10.1080/10511253.2016.1 165854

Miller, J. N. (1993). Tutorial review: Outliers in experimental data and their treatment. *Analyst, 118*(5), 455–461. https://doi.org/10.1039/AN9931800455

Miller, M., Swanson, S. A., & Azrael, D. (2016). Are we missing something pertinent? A bias analysis of unmeasured confounding in the firearm-suicide literature. *Epidemiologic Reviews, 38*(1), 62–69. https://doi.org/10.1093/epirev/mxv011

Mills, L. G., Barocas, B., & Ariel, B. (2012). Restorative justice elements vs. the Duluth model for domestic violence. Experimental Protocol: CRIMPORT. Institute of Criminology, University of Cambridge.

Mills, L. G., Barocas, B., & Ariel, B. (2013). The next generation of court-mandated domestic violence treatment: A comparison study of batterer intervention and restorative justice programs. *Journal of Experimental Criminology, 9*(1), 65–90. https://doi.org/10.1007/s11292-012-9164-x

Mills, L. G., Barocas, B., Butters, R., & Ariel, B. (2015a). The Salt Lake City court-mandated restorative justice treatment for domestic batterers experiment: Part I. Experimental Protocol: CRIMPORT. Institute of Criminology, University of Cambridge.

Mills, L. G., Barocas, B., Butters, R., & Ariel, B. (2015b). The Salt Lake City court-mandated restorative justice treatment for domestic batterers experiment: Part II. Experimental Protocol: CRIMPORT. Institute of Criminology, University of Cambridge.

Mills, L. G., Barocas, B., Butters, R. P., & Ariel, B. (2019). A randomised controlled trial of restorative justice-informed treatment for domestic violence crimes. *Nature Human Behaviour, 3*(12), 1284–1294. https://doi.org/10.1038/s41562-019-0724-1

Minstrell, J. (1982). Explaining the "at rest" condition of an object. *The Physics Teacher, 20*(1), 10–14. https://doi.org/10.1119/1.2340924

Mitchell, R. J. (2017). Frequency versus duration of police patrol visits for reducing crime in hot spots: Non-experimental findings from the Sacramento hot spots experiment. *Cambridge Journal of Evidence-Based Policing, 1*(1), 22–37. https://doi.org/10.1007/s41887-017-0002-2

Mitchell, R. J., Ariel, B., Firpo, M. E., Fraiman, R., del Castillo, F., Hyatt, J. M., Weinborn, C., & Sabo, H. B. (2018). Measuring the effect of body-worn cameras on complaints in Latin America. *Policing, 41*(4), 510–524. https://doi.org/10.1108/PIJPSM-01-2018-0004

Mitchell, R. J., & Lewis, S. (2017). Intention is not method, belief is not evidence, rank is not proof: Ethical policing needs evidence-based decision making. *International Journal of Emergency Services, 6*(3), 188–199. https://doi.org/10.1108/IJES-04-2017-0018

Moher, D., Schulz, K. F., & Altman, D. G. (2001). The CONSORT statement: Revised recommendations for improving the quality of reports of parallel group randomised trials. *BMC Medical Research Methodology, 1*, Article 2. https://doi.org/10.1186/1471-2288-1-2

Moher, D., Shamseer, L., Clarke, M., Ghersi, D., Liberati, A., Petticrew, M., Shekelle, P., Stewart, L. A., & PRISMA-P Group. (2015). Preferred reporting items for systematic review and meta-analysis protocols (PRISMA-P) 2015 statement. *Systematic Reviews, 4*, Article 1. https://doi.org/10.1186/2046-4053-4-1

Monnington-Taylor, E., Bowers, K., Hurle, P. S., Ward, L., Ruda, S., Sweeney, M., Murray, A., & Whitehouse, J. (2019). Testimony at court: A randomised controlled trial investigating the art and science of persuading witnesses and victims to attend trial. *Crime Science, 8*, Article 10. https://doi.org/10.1186/s40163-019-0104-1

Montgomery, P., Grant S., Mayo-Wilson, E., Macdonald, G., Michie, S., Hopewell, S., Moher, D., & CONSORT-SPI Group. (2018). CONSORT-SPI Group. Reporting randomised trials of social and psychological interventions: The CONSORT-SPI, 2018 Extension. *Trials, 19*, Article 407. https://doi.org/10.1186/s13063-018-2733-1

Morgan, S. L., & Winship, C. (2007). *Counterfactuals and causal inference.* Cambridge University Press. https://doi.org/10.1017/CBO9780511804564

Morgan, S. L., & Winship, C. (2012). Bringing context and variability back into causal analysis. In H. Kincaid (Ed.), *Oxford handbook of the philosophy of the social sciences* (pp. 319–354). Oxford University Press. https://doi.org/10.1093/oxfordhb/9780195392753.013.0014

Morgan, S. L., & Winship, C. (2015). *Counterfactuals and causal inference.* Cambridge University Press. https://doi.org/10.1017/CBO9781107587991

Murphy, K., Mazerolle, L., & Bennett, S. (2014). Promoting trust in police: Findings from a randomised experimental field trial of procedural justice policing. *Policing and Society, 24*(4), 405–424. https://doi.org/10.1080/10439463.2013.862246

Murray, A. (2013). Evidence-based policing and integrity. *Translational Criminology,* (5), 4–6.

Murray, D. M. (1998). *Design and analysis of group-randomized trials* (Vol. *29*). Oxford University Press.

Mutz, D., & Pemantle, R. (2012). *The perils of randomisation checks in the analysis of experiments.* ScholarlyCommons. https://repository.upenn.edu/cgi/viewcontent.cgi?article=1767&context=asc_papers

Mutz, D. C., Pemantle, R., & Pham, P. (2019). The perils of balance testing in experimental design: Messy analyses of clean data. *The American Statistician, 73*(1), 32–42. https://doi.org/10.1080/00031305.2017.1322143

Nagin, D. S., Cullen, F. T., & Jonson, C. L. (2009). Imprisonment and re-offending. In M. Tonry (Ed.), *Crime and justice: A review of research* (Vol. *38*, pp. 115–200). University of Chicago Press. https://doi.org/10.1086/599202

Nagin, D. S., & Sampson, R. J. (2019). The real gold standard: measuring counterfactual worlds that matter most to social science and policy. *Annual Review of Criminology, 2*, 123–145. https://doi.org/10.1146/annurev-criminol-011518-024838

Nagin, D. S., Solow, R. M., & Lum, C. (2015). Deterrence, criminal opportunities, and police. *Criminology, 53*(1), 74–100. https://doi.org/10.1111/1745-9125.12057

Nagin, D. S., & Weisburd, D. (2013). Evidence and public policy: The example of evaluation research in policing. *Criminology & Public Policy, 12*(4), 651–679. https://doi.org/10.1111/1745-9133.12030

Near, J. P., & Miceli, M. P. (1985). Organizational dissidence: The case of whistle-blowing. *Journal of Business Ethics, 4*(1), 1–16. https://doi.org/10.1007/BF00382668

Nelson, M. S., Wooditch, A., & Dario, L. M. (2015). Sample size, effect size, and statistical power: A replication study of Weisburd's paradox. *Journal of Experimental Criminology, 11*(1), 141–163. https://doi.org/10.1007/s11292-014-9212-9

Neyman, J. (1923). Próba uzasadnienia zastosowań rachunku prawdopodobieństwa do doświadczeń polowych [On the application of probability theory to agricultural experiments]. *Roczniki Nauk Rolniczych Tom X, 10*, 1–51.

Neyroud, P. W. (2011). *Operation turning point: An experiment in "offender-desistance policing." Experimental Protocol: CRIMPORT*. Institute of Criminology, University of Cambridge. www.criminologysymposium.com/download/18.62fc8fb415c2ea1 06932dcb4/1500042535947/MON11+Peter+Neyroud.pdf

Neyroud, P. W. (2016). The ethics of learning by testing: The police, professionalism and researching the police. In M. Cowburn, L. Gelsthorpe, & A. Wahidin (Eds.), *Research ethics in criminology* (pp. 89–106). Routledge.

Neyroud, P. W. (2017). Learning to field test in policing: Using an analysis of completed randomised controlled trials involving the police to develop a grounded theory on the factors contributing to high levels of treatment integrity in police field experiments [Doctoral dissertation, University of Cambridge]. *Apollo*. https://doi.org/10.17863/CAM.14377

Neyroud, P. W., & Slothower, M. (2015). Wielding the sword of Damocles: The challenges and opportunities in reforming police out-of-court disposals in

England and Wales. In M. Wasik & S. Santatzoglou (Eds.), *The management of change in criminal justice* (pp. 275–292). Palgrave Macmillan. https://doi.org/10.1057/9781137462497_16

Nix, J., & Wolfe, S. E. (2017). The impact of negative publicity on police self-legitimacy. *Justice Quarterly, 34*(1), 84–108. https://doi.org/10.1080/07418825.2015.1102954

Obsuth, I., Cope, A., Sutherland, A., Pilbeam, L., Murray, A. L., & Eisner, M. (2016). London Education and Inclusion Project (LEIP), exploring negative and null effects of a cluster-randomised school-intervention to reduce school exclusion: Findings from protocol-based subgroup analyses. *PLOS ONE, 11*(4), Article e0152423. https://doi.org/10.1371/journal.pone.0152423

Obsuth, I., Murray, A. L., Malti, T., Sulger, P., Ribeaud, D., & Eisner, M. (2017). A non-bipartite propensity score analysis of the effects of teacher–student relationships on adolescent problem and prosocial behavior. *Journal of Youth and Adolescence, 46*(8), 1661–1687. https://doi.org/10.1007/s10964-016-0534-y

Obsuth, I., Sutherland, A., Cope, A., Pilbeam, L., Murray, A. L., & Eisner, M. (2017). London Education and Inclusion Project (LEIP): Results from a cluster-randomised controlled trial of an intervention to reduce school exclusion and antisocial behavior. *Journal of Youth and Adolescence, 46*(3), 538–557. https://doi.org/10.1007/s10964-016-0468-4

Oredsson, S., Jonsson, H., Rognes, J., Lind, L., Göransson, K. E., Ehrenberg, A., … & Farrohknia, N. (2011). A systematic review of triage-related interventions to improve patient flow in emergency departments. *Scandinavian Journal of Trauma, Resuscitation and Emergency Medicine, 19*(1), 43.

Ostle, B., & Malone, L. (2000). *Statistics in research: Basic concepts and techniques for research workers*. Wiley–Blackwell.

Panda, A. (2014). Bringing academic and corporate worlds closer: We need pracademics. *Management and Labour Studies, 39*(2), 140–159. https://doi.org/10.1177/0258042X14558174

Parsons, H. M. (1974). What happened at Hawthorne? New evidence suggests the Hawthorne effect resulted from operant reinforcement contingencies. *Science, 183*(4128), 922–932. https://doi.org/10.1126/science.183.4128.922

Parsons, N. R., Teare, M. D., & Sitch, A. J. (2018). Science forum: Unit of analysis issues in laboratory-based research. *Elife, 7*, Article e32486. https://doi.org/10.7554/eLife.32486

Pashley, N. E., Basse, G. W., & Miratrix, L. W. (2020). Conditional as-if analyses in randomised experiments. arXiv. https://arxiv.org/abs/2008.01029

Payne, J. L. (1974). Fishing expedition probability: The statistics of post hoc hypothesizing. *Polity, 7*(1), 130–138. https://doi.org/10.2307/3234273

Pearl, J. (2010, February). Causal inference. In I. Guyon, D. Janzing, & B. Schölkopf (Eds.), *Proceedings of workshop on causality: Objectives and assessment at NIPS 2008* (Vol. *6*, pp. 39–58). PMLR. http://proceedings.mlr.press/v6/pearl10a/pearl10a.pdf

Pearl, J. (2019). *Causal and counterfactual inference* (Technical Report No. R-485). https://ftp.cs.ucla.edu/pub/stat_ser/r485.pdf

Peduzzi, P., Henderson, W., Hartigan, P., & Lavori, P. (2002). Analysis of randomised controlled trials. *Epidemiologic Reviews*, *24*(1), 26–38. https://doi.org/10.1093/epirev/24.1.26

Pegram, R. (2016). Tracking the implementation of cocooning in preventing near-repeat burglary: A Phase 1 test in Greater Manchester (England) [Unpublished master's thesis]. Institute of Criminology, University of Cambridge.

Perepletchikova, F. (2011). On the topic of treatment integrity. *Clinical Psychology: Science and Practice*, *18*(2), 148–153. https://doi.org/10.1111/j.1468-2850.2011.01246.x

Perepletchikova, F., & Kazdin, A. E. (2005). Treatment integrity and therapeutic change: Issues and research recommendations. *Clinical Psychology: Science and Practice*, *12*(4), 365–383. https://doi.org/10.1093/clipsy.bpi045

Permutt, T. (1990). Testing for imbalance of covariates in controlled experiments. *Statistics in Medicine*, *9*(12), 1455–1462. https://doi.org/10.1002/sim.4780091209

Perry, A. E., Weisburd, D., & Hewitt, C. (2010). Are criminologists describing randomised controlled trials in ways that allow us to assess them? Findings from a sample of crime and justice trials. *Journal of Experimental Criminology*, *6*(3), 245–262. https://doi.org/10.1007/s11292-010-9099-z

Perry, G., Jonathan-Zamir, T., & Weisburd, D. (2017). The effect of paramilitary protest policing on protestors' trust in the police: The case of the "Occupy Israel" movement. *Law & Society Review*, *51*(3), 602–634. https://doi.org/10.1111/lasr.12279

Petrosino, A. J. (1995). Specifying inclusion criteria for a meta-analysis: Lessons and illustrations from a quantitative synthesis of crime reduction experiments. *Evaluation Review*, *19*(3), 274–293. https://journals.sagepub.com/doi/pdf/10.1177/0193841X9501900303

Petrosino, A. J. (2000). How can we respond effectively to juvenile crime? *Pediatrics*, *105*(3), 635–637.

Petrosino, A. J., Boruch, R. F., Soydan, H., Duggan, L., & Sanchez-Meca, J. (2001). Meeting the challenges of evidence-based policy: The Campbell Collaboration. *Annals of the American Academy of Political and Social Science*, *578*(1), 14–34. https://doi.org/10.1177/000271620157800102

Petrosino, A. J., Turpin-Petrosino, C., & Finckenauer, J. O. (2000). Well-meaning programs can have harmful effects! Lessons from experiments of programs such as Scared Straight. *Crime & Delinquency, 46*(3), 354–379. https://doi. org/10.1177/0011128700046003006

Phillips, R. O., Ulleberg, P., & Vaa, T. (2011). Meta-analysis of the effect of road safety campaigns on accidents. *Accident Analysis & Prevention, 43*(3), 1204–1218. https://doi.org/10.1016/j.aap.2011.01.002

Pilotto, A. (2017). Tracking the impact of a randomised controlled trial of burglary investigative methods: A two-year follow up of the Brisbane forensic enhanced examiner response (FEER) time at scene experiment (Australia) [Unpublished master's thesis]. Institute of Criminology, University of Cambridge.

Piquero, A. R., & Brezina, T. (2001). Testing Moffitt's account of adolescence-limited delinquency. *Criminology, 39*(2), 353–370. https://doi. org/10.1111/j.1745-9125.2001.tb00926.x

Piza, E. L., Caplan, J. M., Kennedy, L. W., & Gilchrist, A. M. (2015). The effects of merging proactive CCTV monitoring with directed police patrol: A randomized controlled trial. *Journal of Experimental Criminology, 11*(1), 43–69. https://doi. org/10.1007/s11292-014-9211-x

Piza, E. L., & O'Hara, B. A. (2014). Saturation foot-patrol in a high-violence area: A quasi-experimental evaluation. *Justice Quarterly, 31*(4), 693–718. https://doi.org/1 0.1080/07418825.2012.668923

Piza, E. L., Szkola, J., & Blount-Hill, K. L. (2020). How can embedded criminologists, police pracademics, and crime analysts help increase police-led program evaluations? A survey of authors cited in the evidence-based policing matrix. *Policing.* Advance online publication. https://doi.org/10.1093/police/paaa019

Platz, D. J. (2016). The impact of a value education programme in a police recruit training academy: A randomised controlled trial [Unpublished master's thesis]. University of Cambridge.

Pocock, S. J. (1979). Allocation of patients to treatment in clinical trials. *Biometrics, 35*(1), 183–197. https://doi.org/10.2307/2529944

Pocock, S. J., & Simon, R. (1975). Sequential treatment assignment with balancing for prognostic factors in the controlled clinical trial. *Biometrics, 31*(1), 103–115. https://doi.org/10.2307/2529712

Popper, K. (2005). *The logic of scientific discovery.* Routledge. https://doi. org/10.4324/9780203994627

Porter, S. R., Whitcomb, M. E., & Weitzer, W. H. (2004). Multiple surveys of students and survey fatigue. *New Directions for Institutional Research, 2004*(121), 63–73. https://doi.org/10.1002/ir.101

Raine, A., Portnoy, J., Liu, J., Mahoomed, T., & Hibbeln, J. R. (2015). Reduction in behavior problems with omega-3 supplementation in children aged 8–16 years: A randomised, double-blind, placebo-controlled, stratified, parallel-group trial.

Journal of Child Psychology and Psychiatry, 56(5), 509–520. https://doi.org/10.1111/jcpp.12314

Ratcliffe, J. H., Lattanzio, M., Kikuchi, G., & Thomas, K. (2019). A partially randomized field experiment on the effect of an acoustic gunshot detection system on police incident reports. *Journal of Experimental Criminology, 15*(1), 67–76. https://doi.org/10.1007/s11292-018-9339-1

Ratcliffe, J. H., Taniguchi, T., Groff, E. R., & Wood, J. D. (2011). The Philadelphia foot patrol experiment: A randomised controlled trial of police patrol effectiveness in violent crime hotspots. *Criminology, 49*(3), 795–831. https://doi.org/10.1111/j.1745-9125.2011.00240.x

Ratcliffe, J. H., Taylor, R. B., Askey, A. P., Thomas, K., Grasso, J., Bethel, K. J., Fisher, R., & Koehnlein, J. (2020). The Philadelphia predictive policing experiment. *Journal of Experimental Criminology.* Advance online publication. https://doi.org/10.1007/s11292-019-09400-2

Raudenbush, S. W., & Sampson, R. J. (1999). Ecometrics: Toward a science of assessing ecological settings, with application to the systematic social observation of neighborhoods. *Sociological Methodology, 29*(1), 1–41. https://doi.org/10.1111/0081-1750.00059

Ready, J. T., & Young, J. T. (2015). The impact of on-officer video cameras on police–citizen contacts: Findings from a controlled experiment in Mesa, AZ. *Journal of Experimental Criminology, 11*(3), 445–458. https://doi.org/10.1007/s11292-015-9237-8

Rebers, S., Aaronson, N. K., van Leeuwen, F. E., & Schmidt, M. K. (2016). Exceptions to the rule of informed consent for research with an intervention. *BMC Medical Ethics, 17*, Article 9. https://doi.org/10.1186/s12910-016-0092-6

Ritchie, H. (2019). Gender ratio. Our World in Data. https://ourworldindata.org/gender-ratio

Roberts, C., & Torgerson, D. J. (1999). Baseline imbalance in randomised controlled trials. *British Medical Journal, 319*(7203), 185. https://doi.org/10.1136/bmj.319.7203.185

Rosenbaum, P. R., & Rubin, D. B. (1983). The central role of the propensity score in observational studies for causal effects. *Biometrika, 70*(1), 41–55. https://doi.org/10.1093/biomet/70.1.41

Rosenberger, W., & Lachin, J. M. (2002). *Randomisation in clinical trials: Theory and practice.* Wiley. https://doi.org/10.1002/0471722103

Rosenfeld, R., Deckard, M. J., & Blackburn, E. (2014). The effects of directed patrol and self-initiated enforcement on firearm violence: A randomized controlled study of hot spot policing. *Criminology, 52*(3), 428–449. https://doi.org/10.1111/1745-9125.12043

Rosenthal, R. (1965). The volunteer subject. *Human Relations, 18*(4), 389–406. https://doi.org/10.1177/001872676501800407

Rosenthal, R. (1979). The file drawer problem and tolerance for null results. *Psychological Bulletin, 86*(3), 638–641. https://doi.org/10.1037/0033-2909.86.3.638

Rothwell, P. M. (2005). Subgroup analysis in randomised controlled trials: Importance, indications, and interpretation. *The Lancet, 365*(9454), 176–186. https://doi.org/10.1016/S0140-6736(05)17709-5

Rowlinson, A. (2015). An observational process study of a short program for lower-risk domestic abuse offenders under conditional caution in the Hampshire CARA experiment (England) [Unpublished master's thesis]. Institute of Criminology, University of Cambridge.

Rubin, D. B. (1974). Estimating causal effects of treatments in randomized and nonrandomized studies. *Journal of Educational Psychology, 66*(5), 688–701. https://doi.org/10.1037/h0037350

Rubin, D. B. (1980). Randomization analysis of experimental data: The Fisher randomization test comment. *Journal of the American Statistical Association, 75*(371), 591–593. https://doi.org/10.2307/2287653

Rubin, D. B. (1990a). Comment: Neyman (1923) and causal inference in experiments and observational studies. *Statistical Science, 5*(4), 472–480. https://doi.org/10.1214/ss/1177012032

Rubin, D. B. (1990b). Formal mode of statistical inference for causal effects. *Journal of Statistical Planning and Inference, 25*(3), 279–292. https://doi.org/10.1016/0378-3758(90)90077-8

Rubin, D. B. (2005). Causal inference using potential outcomes. *Journal of the American Statistical Association, 100*(469), 322–331. https://doi.org/10.1198/016214504000001880

Rubin, D. B. (2008). For objective causal inference, design trumps analysis. *Annals of Applied Statistics, 2*(3), 808–840. https://doi.org/10.1214/08-AOAS187

Sahin, N. M. (2014). Legitimacy, procedural justice, and police–citizen encounters: A randomised controlled trial of the impact of procedural justice on citizen perceptions of the police during traffic stops in Turkey [doctoral dissertation, Rutgers University]. *RUcore*. https://doi.org/10.7282/T3CZ35F6

Saint-Mont, U. (2015). Randomisation does not help much, comparability does. *PLOS ONE, 10*(7), Article e0132102. https://doi.org/10.1371/journal.pone.0132102

Salkind, N. J. (2010). *Encyclopedia of research design.* Sage. https://doi.org/10.4135/9781412961288

Salmon, W. C. (1994). Causality without counterfactuals. *Philosophy of Science, 61*(2), 297–312. https://doi.org/10.1086/289801

Sampson, R. J. (2010). Gold standard myths: Observations on the experimental turn in quantitative criminology. *Journal of Quantitative Criminology, 26*(4), 489–500. https://doi.org/10.1007/s10940-010-9117-3

Sampson, R. J., & Laub, J. H. (2003). Life-course desisters? Trajectories of crime among delinquent boys followed to age 70. *Criminology, 41*(3), 555–592. https://doi.org/10.1111/j.1745-9125.2003.tb00997.x

Sampson, R. J., & Laub, J. H. (2017). A general age-graded theory of crime: Lessons learned and the future of life-course criminology. In D. P. Farrington (Ed.), *Integrated developmental and life-course theories of offending* (pp. 175–192). Routledge. https://doi.org/10.4324/9780203788431

Sanetti, L. M. H., & Kratochwill, T. R. (2009). Toward developing a science of treatment integrity: Introduction to the special series. *School Psychology Review, 38*(4), 445–459.

Santos, R. B., & Santos, R. G. (2016). Offender-focused police intervention in residential burglary and theft from vehicle hot spots: A partially blocked randomised control trial. *Journal of Experimental Criminology, 12*(3), 373–402. https://doi.org/10.1007/s11292-016-9268-9

Santos, R. G., & Santos, R. B. (2015). An ex post facto evaluation of tactical police response in residential theft from vehicle micro-time hot spots. *Journal of Quantitative Criminology, 31*(4), 679–698. https://doi.org/10.1007/s10940-015-9248-7

Sargeant, E., Antrobus, E., Murphy, K., Bennett, S., & Mazerolle, L. (2016). Social identity and procedural justice in police encounters with the public: Results from a randomised controlled trial. *Policing and Society, 26*(7), 789–803. https://doi.org/10.1080/10439463.2014.989159

Saunders, J., Hunt, P., & Hollywood, J. S. (2016). Predictions put into practice: A quasi-experimental evaluation of Chicago's predictive policing pilot. *Journal of Experimental Criminology, 12*(3), 347–371. https://doi.org/10.1007/s11292-016-9272-0

Saunders, J., Lundberg, R., Braga, A. A., Ridgeway, G., & Miles, J. (2015). A synthetic control approach to evaluating place-based crime interventions. *Journal of Quantitative Criminology, 31*(3), 413–434. https://doi.org/10.1007/s10940-014-9226-5

Scargle, J. D. (1999). Publication bias (the "file-drawer problem") in scientific inference. arXiv. https://arxiv.org/abs/physics/9909033

Schmidt, F. L. (1992). What do data really mean? Research findings, meta-analysis, and cumulative knowledge in psychology. *American Psychologist, 47*(10), 1173–1181. https://doi.org/10.1037/0003-066X.47.10.1173

Schulz, K. F., & Grimes, D. A. (2002a). Allocation concealment in randomised trials: Defending against deciphering. *The Lancet, 359*(9306), 614–618. https://doi.org/10.1016/S0140-6736(02)07750-4

Schulz, K. F., & Grimes, D. A. (2002b). Generation of allocation sequences in randomised trials: Chance, not choice. *The Lancet, 359*(9305), 515–519. https://doi.org/10.1016/S0140-6736(02)07683-3

Schwartz, C. E., & Sprangers, M. A. (1999). Methodological approaches for assessing response shift in longitudinal health-related quality-of-life research. *Social Science & Medicine, 48*(11), 1531–1548. https://doi.org/10.1016/S0277-9536(99)00047-7

Schwartz, R. D., & Orleans, S. (1967). On legal sanctions. *The University of Chicago Law Review, 34*(2), 274–300. https://doi.org/10.2307/1598934

Schweizer, M. L., Braun, B. I., & Milstone, A. M. (2016). Research methods in healthcare epidemiology and antimicrobial stewardship: Quasi-experimental designs. *Infection Control & Hospital Epidemiology, 37*(10), 1135–1140. https://doi.org/10.1017/ice.2016.117

Scott, W. N., McPherson, G. C., Ramsey, C. R., & Campbell, M. K. (2002). The method of minimization for allocation to clinical trials: A review. *Controlled Clinical Trials, 23*(6), 662–674. https://doi.org/10.1016/S0197-2456(02)00242-8

Senn, S. J. (1989). Covariate imbalance and random allocation in clinical trial. *Statistics in Medicine, 8*(4), 467–475. https://doi.org/10.1002/sim.4780080410

Senn, S. J. (1994). Testing for baseline balance in clinical trials. *Statistics in Medicine, 13*(17), 1715–1726. https://doi.org/10.1002/sim.4780131703

Senn, S. J. (1995). Base logic: Tests of baseline balance in randomised clinical trials. *Clinical Research and Regulatory Affairs, 12*(3), 171–182. https://doi.org/10.3109/10601339509019426

Shadish, W. R. (2013). Propensity score analysis: Promise, reality and irrational exuberance. *Journal of Experimental Criminology, 9*(2), 129–144. https://doi.org/10.1007/s11292-012-9166-8

Shadish, W. R., Cook, T. D., & Campbell, D. T. (2002). *Experimental and quasi-experimental designs for generalized causal inference.* Houghton Mifflin.

Shadish, W. R., & Sullivan, K. J. (2012). *Theories of causation in psychological science.* In H. Cooper, P. M. Camic, D. L. Long, A. T. Panter, D. Rindskopf, & K. J. Sher (Eds.), *APA handbooks in psychology®. APA handbook of research methods in psychology: Vol. 1. Foundations, planning, measures, and psychometrics* (pp. 23–52). American Psychological Association. https://doi.org/10.1037/13619-003

Shaw, M. E. (1981). *Group dynamics: The psychology of small group behavior.* McGraw-Hill.

Shepherd, J. P. (2003). Explaining feast or famine in randomised field trials: Medical science and criminology compared. *Evaluation Review, 27*(3), 290–315. https://doi.org/10.1177/0193841X03027003005

Sherman, L. W. (1997). Communities and crime prevention. In L. W. Sherman, D. C. Gottfredson, D. L. MacKenzie, J. Eck, P. Reuter, & S. Bushway (Eds.), *Preventing crime: What works, what doesn't, what's promising: A report to the United States Congress*. National Institute of Justice. https://citeseerx.ist.psu.edu/viewdoc/download?doi=10.1.1.130.6206&rep=rep1&type=pdf

Sherman, L. W. (1998). *Evidence-based policing*. Police Foundation. http://cebma.org/wp-content/uploads/Sherman-Evidence-Based-Policing.pdf

Sherman, L. W. (2001). Reducing gun violence: What works, what does not, what's promising. *Criminal Justice, 1*(1), 11–25. https://doi.org/10.1177/1466802501001001002

Sherman, L. W. (2003). Misleading evidence and evidence-led policy: Making social science more experimental. *Annals of the American Academy of Political and Social Science, 589*(1), 6–19. https://doi.org/10.1177/0002716203256266

Sherman, L. W. (2007). The power few: Experimental criminology and the reduction of harm. *Journal of Experimental Criminology, 3*(4), 299–321. https://doi.org/10.1007/s11292-007-9044-y

Sherman, L. W. (2009). Evidence and liberty: The promise of experimental criminology. *Criminology & Criminal Justice, 9*(1), 5–28. https://doi.org/10.1177/1748895808099178

Sherman, L. W. (2010). An introduction to experimental criminology. In A. Piquero & D. Weisburd (Eds.), *Handbook of quantitative criminology* (pp. 399–436). Springer. https://doi.org/10.1007/978-0-387-77650-7_20

Sherman, L. W. (2013). The rise of evidence-based policing: Targeting, testing, and tracking. *Crime and Justice, 42*(1), 377–451. https://doi.org/10.1086/670819

Sherman, L. W. (2015). A tipping point for "totally evidenced policing": Ten ideas for building an evidence-based police agency. *International Criminal Justice Review, 25*(1), 11–29. https://doi.org/10.1177/1057567715574372

Sherman, L. W., & Berk, R. A. (1984). The specific deterrent effects of arrest for domestic assault. *American Sociological Review, 49*(2), 261–272. https://doi.org/10.2307/2095575

Sherman, L. W., Bland, M., House, P., & Strang, H. (2016). Targeting family violence reported to Western Australia police, 2010–2015: The felonious few vs. the miscreant many. Report to Deputy Police Commissioner Stephen Brown. Western Australia Police.

Sherman, L. W., & Cohn, E. G. (1989). The impact of research on legal policy: The Minneapolis domestic violence experiment. *Law and Society Review*, 117–144.

Sherman, L. W., Gartin, P. R., & Buerger, M. E. (1989). Hot spots of predatory crime: Routine activities and the criminology of place. *Criminology, 27*(1), 27–56. https://doi.org/10.1111/j.1745-9125.1989.tb00862.x

Sherman, L. W., Gottfredson, D. C., MacKenzie, D. L., Eck, J., Reuter, P., & Bushway, S. D. (1998). *Preventing crime: What works, what doesn't, what's promising.* National Institute of Justice, Office of Justice Programs, U.S. Department of Justice. www. ncjrs.gov/pdffiles/171676.PDF

Sherman, L. W., Neyroud, P. W., & Neyroud, E. (2016). The Cambridge crime harm index: Measuring total harm from crime based on sentencing guidelines. *Policing, 10*(3), 171–183. https://doi.org/10.1093/police/paw003

Sherman, L. W., Schmidt, J. D., & Rogan, D. P. (1992). *Policing domestic violence: Experiments and dilemmas.* Free Press.

Sherman, L. W., Shaw, J. W., & Rogan, D. P. (1995). The Kansas City gun experiment. *Population, 4*, 8–142. https://doi.org/10.1037/e603872007-001

Sherman, L. W., & Strang, H. (2009). Testing for analysts' bias in crime prevention experiments: can we accept Eisner's one-tailed test? *Journal of Experimental Criminology, 5*(2), 185–200.

Sherman, L. W., & Strang, H. (2012). Restorative justice as evidence-based sentencing. *The Oxford handbook of sentencing and corrections* (pp. 215–243). Oxford University Press.

Sherman, L. W., Strang, H., Mayo-Wilson, E., Woods, D. J., & Ariel, B. (2015). Are restorative justice conferences effective in reducing repeat offending? Findings from a Campbell systematic review. *Journal of Quantitative Criminology, 31*(1), 1–24. https://doi.org/10.1007/s10940-014-9222-9

Sherman, L. W., & Weisburd, D. (1995). General deterrent effects of police patrol in crime "hot spots": A randomised, controlled trial. *Justice Quarterly, 12*(4), 625–648. https://doi.org/10.1080/07418829500096221

Sherman, L. W., Williams, S., Ariel, B., Strang, L. R., Wain, N., Slothower, M., & Norton, A. (2014). An integrated theory of hot spots patrol strategy: Implementing prevention by scaling up and feeding back. *Journal of Contemporary Criminal Justice, 30*(2), 95–122. https://doi.org/10.1177/1043986214525082

Shuger, S. L., Barry, V. W., Sui, X., McClain, A., Hand, G. A., Wilcox, S., Meriwether, R. A., Hardin, J. W., & Blair, S. N. (2011). Electronic feedback in a diet-and physical activity-based lifestyle intervention for weight loss: A randomised controlled trial. *International Journal of Behavioral Nutrition and Physical Activity, 8*, Article 41. https://doi.org/10.1186/1479-5868-8-41

Siegel, S. (1957). Nonparametric statistics. *The American Statistician, 11*(3), 13–19. https://doi.org/10.1080/00031305.1957.10501091

Simon, R. (1979). Restricted randomisation designs in clinical trials. *Biometrics, 35*(2), 503–512. https://doi.org/10.2307/2530354

Sindall, K., Sturgis, P., & Jennings, W. (2012). Public confidence in the police: A time-series analysis. *British Journal of Criminology, 52*(4), 744–764. https://doi.org/10.1093/bjc/azs010

Singal, A. G., Higgins, P. D., & Waljee, A. K. (2014). A primer on effectiveness and efficacy trials. *Clinical and Translational Gastroenterology, 5*(1), e45. https://doi.org/10.1038/ctg.2013.13

Skolnick, J. (2002). Corruption and the blue code of silence. *Police Practice and Research, 3*(1), 7–19. https://doi.org/10.1080/15614260290011309

Smith, G. C., & Pell, J. P. (2003). Parachute use to prevent death and major trauma related to gravitational challenge: Systematic review of randomised controlled trials. *British Medical Journal, 327*(7429), 1459–1461. https://doi.org/10.1136/bmj.327.7429.1459

Sobel, M. E. (2006). What do randomised studies of housing mobility demonstrate? Causal inference in the face of interference. *Journal of the American Statistical Association, 101*(476), 1398–1407. https://doi.org/10.1198/016214506000000636

Solomon, R. L. (1949). An extension of control group design. *Psychological Bulletin, 46*(2), 137–150. https://doi.org/10.1037/h0062958

Somers, M. A., Zhu, P., Jacob, R., & Bloom, H. (2013). The validity and precision of the comparative interrupted time series design and the difference-in-difference design in educational evaluation. *MDRC*.

Sousa, W., Ready, J., & Ault, M. (2010). The impact of TASERs on police use-of-force decisions: Findings from a randomised field-training experiment. *Journal of Experimental Criminology, 6*(1), 35–55. https://doi.org/10.1007/s11292-010-9089-1

Spiegelhalter, D. (2019). *The art of statistics: Learning from data*. Penguin.

St. Clair, T., Hallberg, K., & Cook, T. D. (2016). The validity and precision of the comparative interrupted time-series design: Three within-study comparisons. *Journal of Educational and Behavioral Statistics, 41*(3), 269–299. https://doi.org/10.3102/1076998616636854

Steffensmeier, D. J., Allan, E. A., Harer, M. D., & Streifel, C. (1989). Age and the distribution of crime. *American Journal of Sociology, 94*(4), 803–831. https://doi.org/10.1086/229069

Steffensmeier, D. J., Lu, Y., & Na, C. (2020). Age and crime in South Korea: Cross-national challenge to invariance thesis. *Justice Quarterly, 37*(3), 410–435. https://doi.org/10.1080/07418825.2018.1550208

Stern, J. E., & Lomax, K. (1997). Human experimentation. In D. Elliott & J. E. Stern (Eds.), *Research ethics: A reader* (pp. 286–295). University Press of New England.

Stevens, J. P. (2012). *Applied multivariate statistics for the social sciences*. Routledge. https://doi.org/10.4324/9780203843130

Stevens, J. R. (2017). Replicability and reproducibility in comparative psychology. *Frontiers in psychology, 8*, 862.

Stevens, S. S. (Ed.). (1951). *Handbook of experimental psychology*. Wiley.

Stigler, S. M. (1997). Regression towards the mean, historically considered. *Statistical Methods in Medical Research, 6*(2), 103–114. https://doi.org/10.1177/096228029700600202

Strang, H. (2012). Coalitions for a common purpose: Managing relationships in experiments. *Journal of Experimental Criminology, 8*(3), 211–225. https://doi.org/10.1007/s11292-012-9148-x

Strang, H., & Sherman, L. W. (2012). Experimental criminology and restorative justice: Principles of developing and testing innovations in crime policy. In D. Gadd, S. Karstedt, & S. F. Messner (Eds.), *The SAGE handbook of criminological research methods* (pp. 395–410). Sage. https://doi.org/10.4135/9781446268285.n26

Strang, H., Sherman, L., Ariel, B., Chilton, S., Braddock, R., Rowlinson, T., Cornelius, N., Jarman, R., & Weinborn, C. (2017). Reducing the harm of intimate partner violence: Randomised controlled trial of the Hampshire Constabulary CARA Experiment. *Cambridge Journal of Evidence-Based Policing, 1*(2–3), 160–173. https://doi.org/10.1007/s41887-017-0007-x

Strang, H., Sherman, L. W., Mayo-Wilson, E., Woods, D., & Ariel, B. (2013). Restorative justice conferencing (RJC) using face-to-face meetings of offenders and victims: Effects on offender recidivism and victim satisfaction. A systematic review. *Campbell Systematic Reviews, 9*(1), 1–59. https://doi.org/10.4073/csr.2013.12

Stuart, E. A., & Rubin, D. B. (2008). Best practices in quasi-experimental designs. In J. Osborne (Ed.), *Best practices in quantitative methods* (pp. 155–176). Sage. https://doi.org/10.4135/9781412995627.d14

Sullivan, L. E., Fiellin, D. A., & O'Connor, P. G. (2005). The prevalence and impact of alcohol problems in major depression: A systematic review. *American Journal of Medicine, 118*(4), 330–341. https://doi.org/10.1016/j.amjmed.2005.01.007

Sutherland, A., Ariel, B., Farrar, W., & De Anda, R. (2017). Post-experimental follow-ups: Fade-out versus persistence effects: The Rialto police body-worn camera experiment four years on. *Journal of Criminal Justice, 53*, 110–116. https://doi.org/10.1016/j.jcrimjus.2017.09.008

Sutherland, J., & Mueller-Johnson, K. (2019). Evidence vs. professional judgment in ranking "power few" crime targets: A comparative analysis. *Cambridge Journal of Evidence-Based Policing, 3*(1–2), 54–72. https://doi.org/10.1007/s41887-019-00033-z

Sweeten, G., Piquero, A. R., & Steinberg, L. (2013). Age and the explanation of crime, revisited. *Journal of Youth and Adolescence, 42*(6), 921–938. https://doi.org/10.1007/s10964-013-9926-4

Sykes, J. (2015). *Leading and testing body-worn video in an RCT* [Unpublished master's dissertation]. University of Cambridge.

Sytsma, V. A., & Piza, E. L. (2018). The influence of job assignment on community engagement: Bicycle patrol and community-oriented policing. *Police Practice and Research, 19*(4), 347–364. https://doi.org/10.1080/15614263.2017.1364998

Tankebe, J. (2009). Public cooperation with the police in Ghana: Does procedural fairness matter? *Criminology, 47*(4), 1265–1293. https://doi.org/10.1111/j.1745-9125.2009.00175.x

Tankebe, J., & Ariel, B. (2016). *Cynicism towards change: The case of body-worn cameras among police officers* (Hebrew University of Jerusalem Legal Research Paper No. 16-42). SSRN. https://doi.org/10.2139/ssrn.2850743

Taves, D. R. (1974). Minimization: A new method of assigning patients to treatment and control groups. *Clinical Pharmacology & Therapeutics, 15*(5), 443–453. https://doi.org/10.1002/cpt1974155443

Taves, D. R. (2010). The use of minimization in clinical trials. *Contemporary Clinical Trials, 31*(2), 180–184. https://doi.org/10.1016/j.cct.2009.12.005

Taxman, F. S., & Rhodes, A. G. (2010). Multisite trials in criminal justice settings: Trials and tribulations of field experiments. In A. Piquero & D. Weisburd (Eds.), *Handbook of quantitative criminology* (pp. 519–540). Springer. https://doi.org/10.1007/978-0-387-77650-7_25

Taylor, C. (2019, January 28). *What "fail to reject" means in a hypothesis test.* ThoughtCo. www.thoughtco.com/fail-to-reject-in-a-hypothesis-test-3126424

Telep, C. W., Mitchell, R. J., & Weisburd, D. (2014). How much time should the police spend at crime hot spots? Answers from a police agency directed randomised field trial in Sacramento, California. *Justice Quarterly, 31*(5), 905–933. https://doi.org/10.1080/07418825.2012.710645

Thibaut, J. W. (2017). *The social psychology of groups.* Routledge. https://doi.org/10.4324/9781315135007

Thurstone, L. L. (1931). *The reliability and validity of tests: Derivation and interpretation of fundamental formulae concerned with reliability and validity of tests and illustrative problems.* Edwards Brothers. https://doi.org/10.1037/11418-000

Thyer, B. A. (2006). Faith-based programs and the role of empirical research. *Journal of Religion & Spirituality in Social Work: Social Thought, 25*(3–4), 63–82. https://doi.org/10.1300/J377v25n03_05

Toby, J. (1957). Social disorganization and stake in conformity: Complementary factors in the predatory behavior of hoodlums. *Journal of Criminal Law, Criminology & Police Science, 48*(1), 12–17. https://doi.org/10.2307/1140161

Toh, S., & Hernán, M. A. (2008). Causal inference from longitudinal studies with baseline randomisation. *International Journal of Biostatistics*, *4*(1), Article 22. https://doi.org/10.2202/1557-4679.1117

Torgerson, J. D., & Torgerson, C. J. (2003). Avoiding bias in randomised controlled trials in educational research. *British Journal of Educational Studies*, *51*(1), 36–45. https://doi.org/10.1111/1467-8527.t01-2-00223

Travis, L. F., III. (1983). The case study in criminal justice research: Applications to policy analysis. *Criminal Justice Review*, *8*(2), 46–51. https://doi.org/10.1177/073401688300800208

Trowman, R., Dumville, J. C., Torgerson, D. J., & Cranny, G. (2007). The impact of trial baseline imbalances should be considered in systematic reviews: A methodological case study. *Journal of Clinical Epidemiology*, *60*(12), 1229–1233. https://doi.org/10.1016/j.jclinepi.2007.03.014

Trochim, W. M. K. (2006). Internal validity. *Research Methods Knowledge Base*. Available at https://conjointly.com/kb/internal-validity/#:~:text=But%20for%20studies%20that%20assess,is%20perhaps%20the%20primary%20consideration.&text=All%20that%20internal%20validity%20means,%2C%20the%20outcome)%20to%20happen.

Ttofi, M. M., & Farrington, D. P. (2011). Effectiveness of school-based programs to reduce bullying: A systematic and meta-analytic review. *Journal of Experimental Criminology*, *7*(1), 27–56. https://doi.org/10.1007/s11292-010-9109-1

Tyler, T. R., Jackson, J., & Bradford, B. (2014). Procedural justice and cooperation. In G. Bruinsma & D. Weisburd (Eds.), *Encyclopedia of criminology and criminal justice* (pp. 4011–4024). Springer. https://doi.org/10.1007/978-1-4614-5690-2_64

Van Mastrigt, S., Gade, C. B., Strang, H., & Sherman, L. W. (2018). *Restorative justice conferences in Denmark. Experimental Protocol: CRIMPORT*. Institute of Criminology, University of Cambridge. www.crim.cam.ac.uk/documents/KIP-CrimPORT

Vickers, A. J., & Altman, D. G. (2001). Analysing controlled trials with baseline and follow up measurements. *British Medical Journal*, *323*(7321), 1123–1124. https://doi.org/10.1136/bmj.323.7321.1123

Vidal, B. J., & Kirchmaier, T. (2018). The effect of police response time on crime clearance rates. *Review of Economic Studies*, *85*(2), 855–891. https://doi.org/10.1093/restud/rdx044

Villaveces, A., Cummings, P., Espitia, V. E., Koepsell, T. D., McKnight, B., & Kellermann, A. L. (2000). Effect of a ban on carrying firearms on homicide rates in 2 Colombian cities. *JAMA Journal of the American Medical Association*, *283*(9), 1205–1209. https://doi.org/10.1001/jama.283.9.1205

Vollmann, J., & Winau, R. (1996). Informed consent in human experimentation before the Nuremberg code. *British Medical Journal*, *313*(7070), 1445–1447. https://doi.org/10.1136/bmj.313.7070.1445

Volpe, M. R., & Chandler, D. (2001). Resolving and managing conflicts in academic communities: The emerging role of the "pracademic." *Negotiation Journal, 17*(3), 245–255. https://doi.org/10.1111/j.1571-9979.2001.tb00239.x

Von Hofer, H. (2000). Crime statistics as constructs: The case of Swedish rape statistics. *European Journal on Criminal Policy and Research, 8*(1), 77–89. https://doi.org/10.1023/A:1008713631586

Wain, N., & Ariel, B. (2014). Tracking of police patrol. *Policing, 8*(3), 274–283. https://doi.org/10.1093/police/pau017

Wain, N., Ariel, B., & Tankebe, J. (2017). The collateral consequences of GPS-LED supervision in hot spots policing. *Police Practice and Research, 18*(4), 376–390. https://doi.org/10.1080/15614263.2016.1277146

Walker, D. (2010, October 10–13). Being a pracademic: Combining reflective practice with scholarship [Keynote address]. AIPM Conference, Darwin, Northern Territiry, Australia.

Walker, D. H. T., Cicmil, S., Thomas, J., Anbari, F. T., & Bredillet, C. (2008). Collaborative academic/practitioner research in project management: Theory and models. *International Journal of Managing Projects in Business, 1*(1), 17–32. https://doi.org/10.1108/17538370810846397

Walters, G. D., & Bolger, P. C. (2019). Procedural justice perceptions, legitimacy beliefs, and compliance with the law: A meta-analysis. *Journal of experimental Criminology, 15*(3), 341–372.

Webley, P., Lewis, A., & Mackenzie, C. (2001). Commitment among ethical investors: An experimental approach. *Journal of Economic Psychology, 22*(1), 27–42. https://doi.org/10.1016/S0167-4870(00)00035-0

Wei, L., & Zhang, J. (2001). Analysis of data with imbalance in the baseline outcome variable for randomised clinical trials. *Drug Information Journal, 35*(4), 1201–1214. https://doi.org/10.1177/009286150103500417

Weisburd, D. (2000). Randomised experiments in criminal justice policy: Prospects and problems. *Crime & Delinquency, 46*(2), 181–193. https://doi.org/10.1177/0011128700046002003

Weisburd, D. (2003). Ethical practice and evaluation of interventions in crime and justice: The moral imperative for randomised trials. *Evaluation Review, 27*(3), 336–354. https://doi.org/10.1177/0193841X03027003007

Weisburd, D. (2005). Hot spots policing experiments and criminal justice research: Lessons from the field. *Annals of the American Academy of Political and Social Science, 599*(1), 220–245. https://doi.org/10.1177/0002716205274597

Weisburd, D. (2008). *Place-based policing* (Ideas in American Policing Series No. 9). Police Foundation. www.policefoundation.org/wp-content/uploads/2015/06/Weisburd-2008-Place-Based-Policing.pdf

Weisburd, D. (2010). Justifying the use of non-experimental methods and disqualifying the use of randomised controlled trials: Challenging folklore in evaluation research in crime and justice. *Journal of Experimental Criminology, 6*(2), 209–227. https://doi.org/10.1007/s11292-010-9096-2

Weisburd, D. (2015). The law of crime concentration and the criminology of place. *Criminology, 53*(2), 133–157. https://doi.org/10.1111/1745-9125.12070

Weisburd, D., & Britt, C. (2014). *Statistics in criminal justice* (4th ed.). Springer. https://doi.org/10.1007/978-1-4614-9170-5

Weisburd, D., Bushway, S., Lum, C., & Yang, S. M. (2004). Trajectories of crime at places: A longitudinal study of street segments in the city of Seattle. *Criminology, 42*(2), 283–322. https://doi.org/10.1111/j.1745-9125.2004.tb00521.x

Weisburd, D., Cave, B., Nelson, M., White, C., Haviland, A., Ready, J., Lawton, B., & Sikkema, K. (2018). Mean streets and mental health: Depression and post-traumatic stress disorder at crime hot spots. *American Journal of Community Psychology, 61*(3–4), 285–295. https://doi.org/10.1002/ajcp.12232

Weisburd, D., & Eck, J. E. (2004). What can police do to reduce crime, disorder, and fear? *Annals of the American Academy of Political and Social Science, 593*(1), 42–65. https://doi.org/10.1177/0002716203262548

Weisburd, D., Farrington, D. P., & Gill, C. (Eds.). (2016). *What works in crime prevention and rehabilitation: Lessons from systematic reviews*. Springer. https://doi.org/10.1007/978-1-4939-3477-5

Weisburd, D., Farrington, D. P., Gill, C., Ajzenstadt, M., Bennett, T., Bowers, K., Caudy, M. S., Holloway, K., Johnson, S., Lösel, F., Mallender, J., Perry, A., Tang, L. L., Taxman, F., Telep, C., Tierney, R., Ttofi, M. M., Watson, C., Wilson, D. B., & Wooditch, A. (2017). What works in crime prevention and rehabilitation: An assessment of systematic reviews. *Criminology & Public Policy, 16*(2), 415–449. https://doi.org/10.1111/1745-9133.12298

Weisburd, D., & Gill, C. (2014). Block randomised trials at places: Rethinking the limitations of small N experiments. *Journal of Quantitative Criminology, 30*(1), 97–112. https://doi.org/10.1007/s10940-013-9196-z

Weisburd, D., Gill, C., Wooditch, A., Barritt, W., & Murphy, J. (2020). Building collective action at crime hot spots: Findings from a randomised field experiment. *Journal of Experimental Criminology*. Advance online publication. https://doi.org/10.1007/s11292-019-09401-1

Weisburd, D., Groff, E. R., Jones, G., Cave, B., Amendola, K. L., Yang, S. M., & Emison, R. F. (2015). The Dallas patrol management experiment: Can AVL technologies be used to harness unallocated patrol time for crime prevention? *Journal of Experimental Criminology, 11*(3), 367–391. https://doi.org/10.1007/s11292-015-9234-y

Weisburd, D., Groff, E. R., & Yang, S. M. (2012). *The criminology of place: Street segments and our understanding of the crime problem.* Oxford University Press. https://doi.org/10.1093/acprof:oso/9780195369083.001.0001

Weisburd, D., Lum, C. M., & Petrosino, A. (2001). Does research design affect study outcomes in criminal justice? *Annals of the American Academy of Political and Social Science, 578*(1), 50–70. https://doi.org/10.1177/000271620157800104

Weisburd, D., Lum, C. M., & Yang, S. M. (2003). When can we conclude that treatments or programs "don't work"? *Annals of the American Academy of Political and Social Science, 587*(1), 31–48. https://doi.org/10.1177/0002716202250782

Weisburd, D., Petrosino, A., & Mason, G. (1993). Design sensitivity in criminal justice experiments. *Crime and Justice, 17*, 337–379. https://doi.org/10.1086/449216

Weisburd, D., & Taxman, F. S. (2000). Developing a multicenter randomised trial in criminology: The case of HIDTA. *Journal of Quantitative Criminology, 16*(3), 315–340. https://doi.org/10.1023/A:1007574906103

Welsh, B. C., Braga, A. A., & Bruinsma, G. J. (Eds.). (2013). *Experimental criminology: Prospects for advancing science and public policy.* Cambridge University Press. https://doi.org/10.1017/CBO9781139424776

Welsh, B. C., & Farrington, D. P. (2001). Toward an evidence-based approach to preventing crime. *Annals of the American Academy of Political and Social Science, 578*(1), 158–173. https://doi.org/10.1177/000271620157800110

Welsh, B. C., & Farrington, D. P. (2012). Crime prevention and public policy. In D. P. Farrington & B. C. Welsh (Eds.), *The Oxford handbook of crime prevention* (pp. 3–19). Oxford University Press. https://doi.org/10.1093/oxfordhb/9780195398823.013.0001

Welsh, B. C., Podolsky, S. H., & Zane, S. N. (2020). Between medicine and criminology: Richard Cabot's contribution to the design of experimental evaluations of social interventions in the late 1930s. *James Lind Library Bulletin: Commentaries on the History of Treatment Evaluation.* www.jameslindlibrary.org/articles/between-medicine-and-criminology-richard-cabots-contribution-to-the-design-of-experimental-evaluations-of-social-interventions-in-the-late-1930s/

White, H. (2006). *Impact evaluation: The experience of the independent evaluation group of the World Bank.* The World Bank.

Whitehead, T. N. (1938). *The industrial worker.* Harvard University Press.

Wicherts, J. M., Veldkamp, C. L., Augusteijn, H. E., Bakker, M., Van Aert, R., & Van Assen, M. A. (2016). Degrees of freedom in planning, running, analyzing, and reporting psychological studies: A checklist to avoid p-hacking. *Frontiers in Psychology, 7*, Article 1832. https://doi.org/10.3389/fpsyg.2016.01832

Wikström, P.-O. H. (2008). In search of causes and explanations of crime. In R. King, & E. Wincup (Eds.), *Doing research on crime and justice* (pp. 117–140). Oxford University Press.

Wikström, P.-O. H. (2010). Explaining crime as moral actions. In S. Hitlin & S. Vaisey (Eds.), *Handbook of the sociology of morality* (pp. 211–239). Springer. https://doi.org/10.1007/978-1-4419-6896-8_12

Wilcoxon, F. (1992). Individual comparisons by ranking methods. In S. Kotz & N. L. Johnson (Eds.), *Breakthroughs in statistics* (pp. 196–202). Springer. https://doi.org/10.1007/978-1-4612-4380-9_16

Willis, J. J. (2016). The romance of police pracademics. *Policing, 10*(3), 315–321. https://doi.org/10.1093/police/paw030

Willson, V. L., & Putnam, R. R. (1982). A meta-analysis of pretest sensitization effects in experimental design. *American Educational Research Journal, 19*(2), 249–258. https://doi.org/10.3102/00028312019002249

Wilson, M. (2013). The Sycamore Tree victim awareness programme for released prisoners. Experimental Protocol: CrimPORT. Institute of Criminology, University of Cambridge. www.crim.cam.ac.uk/global/docs/wilson2013.pdf

Young, J. (1980). Thinking seriously about crime. In M. Fitzgerald, G. McLennan, & J. Pawson (Eds.), *Crime and society: Readings in history and theory* (pp. 248–309). Routledge.

Young, J. (2014). Implementation of a randomised controlled trial in Ventura, California: A body-worn-video camera experiment (United States) [Unpublished master's thesis]. Institute of Criminology, University of Cambridge.

Yusuf, S., Collins, R., & Peto, R. (1984). Why do we need some large, simple randomised trials? *Statistics in Medicine, 3*(4), 409–420. https://doi.org/10.1002/sim.4780030421

INDEX